ROUTLEDGE LIBRARY EDITIONS: HISTORY
AND PHILOSOPHY OF SCIENCE

THE SOCIAL IMPACT OF
MODERN BIOLOGY

THE SOCIAL IMPACT OF MODERN BIOLOGY

Edited by

WATSON FULLER

Volume 12

LONDON AND NEW YORK

First published in 1971
This edition first published in 2009 by
Routledge
2 Park Square, Milton Park, Abingdon, Oxon, OX14 4RN
Simultaneously published in the USA and Canada
by Routledge
711 Third Avenue, New York, NY 10017
Routledge is an imprint of the Taylor & Francis Group, an informa business

First issued in paperback 2011

© 1971 The British Society for Social Responsibility in Science and Watson Fuller

All rights reserved. No part of this book may be reprinted or reproduced or utilised in any form or by any electronic, mechanical, or other means, now known or hereafter invented, including photocopying and recording, or in any information storage or retrieval system, without permission in writing from the publishers.

British Library Cataloguing in Publication Data
A catalogue record for this book is available from the British Library

Library of Congress Cataloging in Publication Data
A catalog record for this book has been requested

ISBN 13: 978-0-415-42029-7 (Set)
ISBN 13: 978-0-415-44091-2 (Volume 12) (hbk)
ISBN 13: 978-0-415-61179-4 (Volume 12) (pbk)

Publisher's Note
The Publisher has gone to great lengths to ensure the quality of this reprint but points out that some imperfections in the original copies may be apparent.

Disclaimer
The Publishers have made every effort to trace copyright holders and would welcome correspondence from those they have been unable to contact.

The social impact of modern biology

Edited by

Watson Fuller
Reader in Biophysics, King's College
University of London

Routledge & Kegan Paul
London

First published 1971
by Routledge & Kegan Paul Ltd
Broadway House, 68–74 Carter Lane,
London EC4V 5EL
Printed in Great Britain by
Cox & Wyman Ltd,
London, Reading and Fakenham
© The British Society for Social Responsibility in Science
and Watson Fuller 1971
No part of this book may be reproduced in
any form without permission from the
publisher, except for the quotation of brief
passages in criticism

ISBN 0 7100 6676 7(c)
     0 7100 7054 3(p)

Contents

	Preface	1
Part one	**Science, technology and values**	
One	Introduction M. H. F. Wilkins	5
Two	On the logical relationship between knowledge and values Jacques Monod	11
Three	Fragmentation in science and in society David Bohm	22
Part Two	**Molecular genetics**	
Four	Molecular genetics: an introductory background W. Hayes	39
Five	Molecular genetics: short-term applications and long-term possibilities M. R. Pollock	60
Part three	**Human genetics and reproduction**	
Six	Social effects of research in human genetics Geoffrey Beale	84

Seven	**Ethics and eugenics** L. S. Penrose	94
Eight	**The obstetrician and genetic investigation** D. V. I. Fairweather	102
Nine	**Aspects of human reproduction** R. G. Edwards	108
Part four	**Immunology and cancer**	
Ten	**Some implications of modern immunology** J. H. Humphrey	122
Eleven	**Molecular biological approach to the cancer problem** J. D. Watson	140
Part five	**Agricultural botany and the environment**	
Twelve	**Molecular biology and agricultural botany** Arthur W. Galston	154
Thirteen	**Environmental problems and the reunification of the scientific community** Joseph G. Hancock	167
Part six	**Science in society**	
Fourteen	**The role of industry in applied science** A. J. Hale	175
Fifteen	**Science and social values** Martin M. Kaplan	192
Sixteen	**Evolutionary biology and ideology: Then and now** Robert M. Young	199

Seventeen	**The myth of the neutrality of science** Steven Rose and Hilary Rose	215
Eighteen	**The scientist in opposition in the United States** Jon Beckwith	225
Nineteen	**The disestablishment of science** J. Bronowski	233
Twenty	**Possible ways to rebuild science** M. H. F. Wilkins	247
	Contributors to the discussion	255

The contributors

M. H. F. Wilkins, FRS. Nobel Laureate (1962) for determination of the structure of DNA.
Jacques Monod, For.Mem.RS. Nobel Laureate (1965) for work on mechanisms controlling protein synthesis. Author of *Le Hasard et la Nécessité* (1970).
David Bohm. Author of *Quantum Theory, Causality and Chance in Modern Physics* and *Special Theory of Relativity*.
W. Hayes, FRS. Author of *The Genetics of Bacteria and their Viruses*.
M. R. Pollock, FRS.
Geoffrey Beale, FRS. Author of *The Genetics of Paramecium Aurelia*.
L. S. Penrose, FRS. Author of *The Biology of Mental Defect* and *The Objective Study of Crowd Behaviour*.
D. V. I. Fairweather.
R. G. Edwards.
J. H. Humphrey, FRS. Author (with R. G. White) of *Immunology for Students of Medicine*.
Arthur W. Galston. Author of *The Life of the Green Plant*, (with J. Bonner) *Principles of Plant Physiology* and (with P. J. Davies) *Control Mechanisms in Plant Development*.
J. D. Watson. Nobel Laureate (1962) for work on the structure of deoxyribonucleic acid. Author of *The Double Helix* and *The Molecular Biology of the Gene*.
Joseph G. Hancock.
A. J. Hale.
Martin M. Kaplan.
Robert M. Young. Author of *Mind, Brain and Adaptation in the Nineteenth Century*.
Jon Beckwith. Editor (with D. Zipser) of *The Lactose Operon*.
Hilary Rose. Author (with Steven Rose) of *Science and Society* and *The Housing Problem*.
Steven Rose. Author of *The Chemistry of Life* and editor of *Chemical and Biological Warfare*.
J. Bronowski. Author of *The Common Sense of Science, Science and Human Values* and *The Identity of Man*.

Preface

Within the next few decades discoveries in biology can radically change human life as we know it. While it is true that biology has already had a substantial impact on medicine and food production, by comparison to future possibilities these effects are piecemeal and trivial. Advances in the treatment of disease, in understanding the processes of ageing, in the control of behaviour, in the harnessing of micro-organisms to produce particular chemicals and in food production have potentialities which will expose the limitations of the social and political structures which have evolved for the application of scientific and technical advances. The problem is not simply that we may not derive the maximum benefit from these advances but that if this new information is not correctly applied, our well-being or even the survival of our species is threatened. We are threatened physically as a species by the depletion of resources and poisoning of our environment because of unplanned technological developments. We are threatened socially as a species by the techniques for the control of behaviour which could dehumanize and could destroy creativity. As individuals, our value systems are being undermined as new knowledge on biological phenomena such as heredity and behaviour undermines the myths and dogmas at the heart of our ethical and religious beliefs.

The British Society for Social Responsibility in Science felt that a meeting to discuss these issues could make a real contribution to defining the problems which can be expected to arise and to the development of attitudes and policies to help solve these problems. We attempted to achieve this in the following ways: first, by providing authoritative analyses of likely developments in areas such as genetics, cancer, immunology and agricultural botany by scientists actively working in these areas;

second, by attempting to assess the likely social impact of these developments by inviting contributions from scientists who had thought particularly about these problems and also from non-scientists; third, by programming the meeting so that there was ample opportunity not only for the speakers but also for the audience to have a full discussion on these topics.

From the first the meeting was planned to be public and the press and broadcasting organizations invited. However, it was recognized that there would be very many people not able to attend who would wish to know more about the meeting than could be reported by the media and further that even those present would welcome the opportunity to study the material more intensively. The talks and discussion have therefore been edited to make this book. All discussion speakers were sent my preliminary editing of their contribution and their replies have been most helpful in the final editing. For reasons of space, the discussion has been rather heavily pruned; also I have rearranged contributions to group those on related topics. I have not attempted to impose any uniformity of style on the manuscripts supplied by speakers and have done no more than attempt to minimize overlap. They are diverse not only in content but also in approach and attitude and are, I feel, as interesting from the point of view of the attitudes they indicate in people in a wide range of positions, as from what they actually say.

The general form of the conference was determined by Maurice Wilkins, who made initial contact with most of the speakers. Advice came from many scientists, especially from Arthur Galston, Henry Harris, John Humphrey and Martin Pollock, and from the British Pugwash Group. In addition to the great debt we owe to the speakers we wish to thank the chairman, Dr C. O. Carter, Professor W. R. S. Doll, Dr M. F. Perutz and Professor Pat Wall. Clearly I owe a great debt to all the contributors for the co-operation and promptness which allowed the complete manuscript to be in the hands of the printers less than seven weeks from the end of the meeting. For this aspect I must also record my considerable debt to John Wolfers and Ian Roxburgh for their very helpful professional advice, to Bert Ede and the BBC who recorded the proceedings, to my wife who transcribed the discussion, and to my colleagues at King's College and, in particular Ann Kernaghan, for all their help. Very effective organizational support for the meeting was provided by David Dickson and the BSSRS administrative staff. Robert Avery and the staff at Friends' House provided such excellent meeting facilities. The meeting would hardly have been possible without an early guarantee of financial support from the Council for Biology in Human Affairs, Salk Institute, San Diego, California.

Finally I should record my debt to Matteo Adinolfi, Geoffrey Brown, David Dickson, Shev Lal, Steven Rose, Bob Smith and Maurice Wilkins, my colleagues on the meeting organizing committee, which held regular meetings during the eight-month period over which the detailed programme was planned. The success of the meeting, at which we had an attendance throughout of over 800, bears ample testimony to their contribution. I hope all who helped to make the meeting possible will feel that this book provides an adequate record and useful basis for further study.

Watson Fuller

Part one Science, technology and values

One Introduction
M. H. F. Wilkins
Professor of Biophysics; Director, Medical Research Council Unit, Biophysics Laboratories, King's College, University of London, President of BSSRS

The crisis in science today has not only direct bearing on the question of our survival but is of deep significance in relation to our fundamental beliefs and our value-judgments. Some scientists, of course, would deny that there *is* a crisis. They continue their work much as scientists have done for the past half-century. But the great appreciation of the tranquillity in British laboratories shown by visiting scientists from the USA and from Japan is a sharp reminder of how different things are there. In the USA, in one university, a research building was blown up and, in others, work of famous laboratories was seriously hampered for months on end; in Japan, where science has been forging ahead, disruption is now widespread. While much of this disturbance arises from general student unrest and political frustration, a considerable amount of it is directly concerned with science, with its organization and with its social priorities.

Yet even where scientists continue to work undisturbed, their attitude to their work has, since the war, significantly changed. Although many scientists regard their work unquestioningly, in general there has been a perceptible change. The main cause is probably the Bomb: scientists no longer have their almost arrogant confidence in the value of science. At the same time non-scientists increasingly and openly question the value of science. There are extremists who go further and object to rational thought as a whole. Dimly perceiving that reason alone does not comprehend experience in its entirety, they renounce reason and advocate mysticism as the sole guide through our lives and into the future. The breakdown of confidence in reason is illustrated by the popularity of such things as debased forms of astrology. That this anti-rational attitude is probably spreading is indicated by its prevalence among younger people, including intellectuals. Significantly, although science is increasingly involved in modern living, schoolchildren are choosing to study other subjects. It is alarming to remember that civilized man has only a thin crust of rationality and that not long ago, in one of the most highly civilized countries of the West, books were banned, even publicly burnt, and genocide, based on a 'mystical' ideology, was government policy. Unfortunately, such things do not belong only to the past.

Preoccupation with the Bomb and pollution causes the value of science to be questioned. But leaving aside what is plainly negative and destructive, the extraordinary expenditure on getting men to the moon and the creation of a population problem as a result of the humane application of science in medicine make it no longer possible to regard science as automatically providing benefits if suitably funded. Things were never so simple as that, although those in government may have thought them so. The results of science were always rather unpredictable; but today, more than ever before, there are complications, undesired side-effects and difficult choices involving value-judgments. Furthermore, people are disturbed by the rapidity with which science is giving rise to changes in our living: the motorcar, pill, computer, and so on, insidiously change our attitudes. Rapidity of change makes it difficult to learn from experience. Science is seen as dehumanizing: the Brave New World now almost reality, babies in test-tubes, men replaced by computers or regarded as computers, logical thought driving out human warmth and poetry. The radical Left regards science more as a repressive tool of the establishment than as a force for improving and changing society.

Scientists have various reactions to these difficulties. Most scientists have a vague entrenched commitment but are less articulate than their critics. Many regard discussion of difficulties (by, they say, the Jeremiahs of science) as rocking the boat and inevitably leading to over-emphasis of the less desirable side. They condemn alarmist prophecies of doom, and rightly so, because these cause over-reaction, which either takes the form of overall condemnation of science or of putting one's head in the sand. Scientists also rather naturally resent those who lecture them about the need to be more socially responsible. Their instinct is probably correct that little progress will come from merely adding to the body of science a layer of social responsibility. They are doubtless right, too, in feeling that the rather out-of-date idea that science needs maximum freedom to develop wherever it will has, in some peculiar way, great value. Nevertheless, nearly all scientists concede that the urgency of man's situation is so great that something must be done.

In all these considerations the first thing is to get an adequate idea of the value of science in the world today. One of the main ideas behind this conference was that new ideas about this would emerge. Need for a sense of value is not peculiar to science, but pertains to higher learning generally. To say that science has value because its application can bring man great benefits is true, but it is not enough. The benefits are largely material, such as improving the standard of living or conquering diseases. These are of the highest priority in backward countries, but in technologically advanced countries need for material advance is much less urgent and young people especially tend to express more interest in the quality of life in terms of spiritual values and human relationships.

Clearly we must establish the value of science in more than material terms. For the 17th century this has been well described by Medawar in his Presidential Address to the British Association in 1969. The development of experimental science in the 17th century swept away the negative dogmatic influence of many centuries of scholasticism, and opened the way to the Age of Reason and the French Revolution. The liberation of man's spirit was symbolized by the opening up of the New World.

It is still possible today to catch some of that imaginative uplift. Consider for example the branch of science that deals with nervous systems. Such science should not only lead to control of nervous disorders but, by providing understanding of how the human brain works, should throw new light on the nature of the mind itself. This understanding should (to use hippie language) *expand the mind*. Darwin and Freud greatly altered man's attitude towards himself and his sense of values, and as a result we can now take into account our animal nature and our psyche and we have new bases for tolerance between men. As neurobiology develops, e.g. when neurophysiology and psychology interpenetrate effectively, we may expect further illumination of our nature, and such self-knowledge should greatly influence our values. Science is valuable then, in terms of the self-knowledge that it gives. Since we are part of the material world, we must study the whole of that world in order to understand ourselves more fully. Therein, I believe, lies the value of all science.

We must, however, accept that today we cannot recover the confidence and uplift that science gave man in the past. Although the prospects of, for example, greater self-knowledge through neuroscience are exciting, we know that self-knowledge does not necessarily give rise to wisdom and that there is fear that neuroscience will be misapplied in mind-control. We are oppressed by our limitations; we know in the case of the Bomb that it is fear rather than generosity or tolerance that has restrained us from frightful acts.

The situation is full of difficulties. It may be easier if we try to think about our problems differently. By posing the problem as one of relating knowledge, wisdom and human needs, we imply that knowledge, wisdom and needs are separated and distinct. Such fragmentation exists not only in science but also in our thinking about it. It has not always existed. In the 17th century, for instance, the concepts of pure and applied science did not exist: human needs, whether practical or philosophical, were part and parcel of science. During the past three hundred years the concept of pure science developed, and it was no doubt necessary in order to enable specialization to accelerate the growth of science, but our present difficulties would be reduced if we now abandoned that concept.

What needs does science satisfy? The ability to 'do science' presumably evolved out of the problem-solving ability that man acquired, through natural selection, to meet the needs of survival. If we accept that man has a basic need for continuing mental development – if he does not develop he will regress – this need is itself related to survival. In that way all basic human needs are linked together.

To establish more clearly the value of science, it is necessary that we should no longer have separate compartments for pure and for applied science, human needs and so on. Scientific research and teaching should not be remote from living as a whole (meaning the individual's emotional and intellectual development, cultural activities, social organization). Lord Snow has seen Two Cultures separated in two groups of men – of the arts and of science. What is more serious is that the separation exists within the individual. A way of viewing fragmentation in science derives from Jung's idea that problems of individual psychology are to be solved by bringing together the thinking and the feeling parts of man. The concept of social responsibility in science implies that science too has its thinking and feeling parts.

If these parts could be brought together, science would be better related to man's wider hopes and needs: dehumanizing aspects of science would be reduced, science would be a force for changing and improving society and social responsibility would be implicit in the nature of science itself.

These ideas may be made clearer by exploring the relation of science and art. There are common problems in these two areas. The crisis in science is only part of a larger cultural crisis. Art, higher learning generally, politics, economics – none of these activities today is properly related to human needs: all are fragmented. Like science, art has become remote from living: it is not, as it has been at times for instance in primitive societies, an integral part of our culture, existing in all aspects of life, in everyday matters, in religion. Art has become separate and specialized, understood only by a minority, segregated in galleries, museums and concert halls (and, like science, exploited for political and commercial ends). New kinds of activity in the art-science area, e.g. ecology lectures and art/technology exhibitions at the Institute of Contemporary Arts here in London, suggest changes in the wind, the possibility that culture may be re-establishing itself on a wider basis.

Adopting the ideas that I have endeavoured to express should lead to changes in science itself. There are indeed signs of such changes. The demand so often made by radical students (and often appearing superficial) for 'socially relevant' science may have already produced some desirable change. This is mainly so in the USA. New research projects on environmental problems, sometimes incorporated in undergraduate teaching, have raised new kinds of intellectual challenge. Academic research workers pushed into social relevancy have found themselves, to their surprise, doing new work of great fundamental, as well as practical, interest. These developments do not simply mean that new areas of applied research are being created, neither do they mean that interest in fundamental ideas need be lost. Rather, it may be that by becoming more adequately related to human needs in the widest sense, science becomes of *greater* fundamental interest.

It is sometimes supposed that desire to avoid dangerous applications of science might lead to society imposing undesirable controls on scientific research: because there is an essential element of unpredictability in science, research which would ultimately be seen as very desirable might be blocked. For example, the Church presumably believed it was being socially responsible when it tried to stop Galileo. However, this kind of situation should not be serious if science and society are not fragmented. The need for a basic element of freedom in scientific enquiry would be recognized by an unfragmented society as a very important human need which must be set alongside other human needs.

In the 17th century there was a very critical phase in the development of science. This was a positive phase with confidence in science. Today, after much development of science over 300 years, there is for the first time, another very critical phase; but this time it is negative and there is loss of confidence in science. To deal with this crisis we can expect to need in science new approaches as radical as those developed in the critical phase of the 17th century. In considering this problem, thinking may not get us far unless we adopt the experimental approach (*nullius in verba*, as the founders of the Royal Society put it in over-emphatic reaction against scholasticism). Setting up new research projects, devising new teaching courses, organizing discussions, are all experiments that may produce unpredicted results. We begin our present discussions with attempts to build intellectual frameworks and with discussion of the value of science. We then consider various aspects of biology and their applications. We cannot consider all aspects, that would be much too large a task. We start from microbial genetics, proceed to human genetics, and then pass on to various aspects of molecular biology, cell biology and medicine. We necessarily become involved with medical ethics. We consider the various fields not so much because we are interested specifically in their details – though considering the details may be necessary for our purpose – but because we are interested in what these considerations may reveal about the general principles of science and society. In the talks on various fields of molecular biology we hear about the very positive contributions biology makes, for instance in developing new crops and dealing with the cancer problem. We consider lessons which can be learnt from the history of science and also from industrial science. Towards the end of this book we will be concerned with philosophical and political problems – because these now obtrude themselves increasingly on scientists. It is difficult today for a scientist to carry out meaningful work remote from the world outside and remote from philosophy and politics.

Two On the logical relationship between knowledge and values

Jacques Monod
Head of Department of Cellular Biochemistry, Pasteur Institute,
Professor of Molecular Biology,
Collège de France, Paris

No one would deny that in this 'age of science' when society lives and depends on its technology (and therefore ultimately on science itself) as much, or more, than does a 'junkie' on his favourite drug, the scientist does bear a heavy responsibility. Before even attempting to clarify the nature and extent of this responsibility, and of the obligations which it entails, it might be well to consider briefly some preliminary questions regarding the status of science, that is, of *objective knowledge* in relation to ethics.

Responsibilities, duties, obligations can be rationally defined and discussed only in respect to an accepted system of values. True enough there are many problems, or alternatives, where a choice can be made, a judgment may be passed, without going to the very foundations of our value systems. Everyone will agree that pollution is a BAD THING, and therefore anti-pollution techniques or efforts, petitions, protests, warnings are GOOD. Everyone also agrees that genetic defects are BAD. But as we know only too well there are wide divergences of opinion as at which technically feasible methods may be considered ethically acceptable as part of an effort to prevent the gradual genetic deterioration of the human species. Many still draw a line at therapeutic abortion. Others who will recommend therapeutic abortion would violently oppose any sort of eugenic legislation. Ninety per cent of the international scientific community agree that war is an absolutely BAD THING. Therefore a large proportion of these 90 per cent will systematically refuse to have anything to do, as scientists, in the development of new weapons or 'systems' in their own country. The question is: what would they do, were they now confronted with the same dilemma that was met and resolved by Szilard, Fermi and Einstein in 1940, when they took it upon themselves to persuade Roosevelt to start what was later known as the 'Manhattan Project'? My personal guess is that most of these well-meaning scientists (even including Linus Pauling himself) would, under similar circumstances, act exactly as Einstein did.

Quite obviously, to choose one or another course of action when in the presence of such alternatives amounts not to a judgment of knowledge, but to one of values. Objective knowledge defines the alternatives. By itself it does not, in any way, help in solving the agonizing ethical problem which science only poses. It is hardly necessary to point out that deciding to abstain and do nothing about such problems also amounts to a choice. Certainly not the most respectable one.

Science rests upon a strictly *objective* approach to the analysis and interpretation of the universe, including Man himself and human societies. Science ignores, and must ignore, value judgments. Knowledge yet discloses and inevitably suggests new possibilities of action. But to *decide* upon a course of action is to step out of the realm of objectivity into that of values which, by essence, are non-objective and therefore cannot be derived from objective knowledge. There is strictly no way of *objectively* proving that it is BAD to make war, or kill a man, or to rob him, or to sleep with one's own mother.

It might appear therefore that a scientist, defined as professionally devoted exclusively to the pursuit of objective knowledge, does *not* bear any *special* responsibility towards society, or human welfare. Indeed it is not clear that, in virtue of his training, he is any better equipped than a theologian, a lawyer, a policeman or a professional politician, to recommend or suggest any particular choice in a meaningful course of action.

It might be suggested that the only *obligation* inherent to his status is at least to *inform* society of the objective state of affairs pertaining to his field, and of the probable consequences of various decisions that might be taken. Even that obligation is doubtful, however. Suppose you, as a scientist, had discovered a simple, effective and cheap way of killing any particular segment of humanity that might prove a nuisance to others. Is it your duty to disclose your discovery to the world? or only to your own government? or to the general secretary of your party? or as probably most scientists would, only to the scientific community, albeit in terms so technical that only scientists could understand the implications of your findings? This last 'course of action' would assure both your fame within the community and the certainty that your discovery would inevitably, one day, become public property and a threat to mankind. Will you do that, as indeed some nuclear scientists did in the late thirties, when they held such a threat in their hands, and knew they did?

Consider yet another problem. Suppose you were a social anthropologist also competent in experimental psychology and in genetics. Would you not be tempted to use this (rare indeed) combination of talents to study the genetics of intelligence, for instance? And suppose, further, that you had at last devised truly 'culture-free' tests that would enable you to compare the average genetic potential of different human races. By the time you had reached this point you should know that you are touching upon an emotionally and politically very dangerous subject. Will you stop at this point, for fear of perhaps discovering unpleasant facts? Or go ahead, in due respect to the supposedly 'absolute' value of knowledge? Suppose you did, and you did find that certain ethnical groups, in your own country, are indeed definitely 'inferior' in respect to certain standards? What now will you do with the data? Publish them even though you knew what abuse could be made of them? Or keep them to yourself as a dark secret, thus violating one of the basic moral laws of the scientific community, namely that new knowledge belongs not to the individual, but to the community, and therefore must not be concealed?

I won't attempt to answer the question here. I suggest you might try, just as an exercise in the ethics of science.

Now, it is my feeling that, in spite of these extreme 'casuistic' difficulties, scientists do bear a very special responsibility to society and mankind as a whole, and also that within precisely defined limits they are, or should be, better equipped through their professional advocation and training to make valid recommendations in the choice between certain alternatives which may engage the future of mankind. In the following I shall try to justify this feeling on the basis of a purely *logical* argument.

Schematically, the argument runs as follows (some of the intervening statements are such obvious matters of fact that I shall not try to justify or illustrate them):

1. All the traditional systems of value (including those that our societies still pretend to live by) were based on some kind of 'history' or ontogeny, that presented men with absolute moral laws which it was beyond their powers to discuss, dismiss or reject. The ethical system did not belong to Man: he belonged to

it. This applies to religious systems of course, but equally to the systems stemming from the 'enlightment' which assumed the existence of 'natural' moral laws. Examples: the ethical bases of the American and French revolutions, and the Marxian belief in 'objective' laws of history.

2. The scientific approach has destroyed not the values recommended by these systems (science is no judge of values) but the various mythical histories or philosophical ontogenies upon which these systems were believed to be firmly based.

3. Hence, modern societies, living both economically and psychologically upon the technological fruits of science, have been robbed, by science itself, of any firm, coherent, acceptable 'belief' upon which to base their value systems. This, probably, is the greatest revolution that ever occurred in human culture. I mean, again, the utter destruction, by science, by the systematic pursuit of objective knowledge, of all the belief systems, whether primitive or highly sophisticated, which had, for thousands of years, served the essential function of justifying the moral code and the social structure.

4. This revolution is at the very root of the modern *mal du siècle*. Whether clearly or only dimly perceived in its nature and consequences, it is the cause of the profound feeling of frustration and alienation which pervades contemporary men, especially the young. No wonder this frustration often expresses itself by a systematic, at times violent, rejection not only of the technological 'gifts' of science, but of science itself as a pursuit, and of objectivity as a *moral* attitude.

5. It is at this historical and logical point that the social responsibility of science is inevitably and most deeply engaged. Of science, and therefore of the scientists themselves as representatives and heralds of their discipline. Clearly, it is for them to resolve, if it can be resolved at all, this most profound of all the dilemmas inherent in the relationship between science and society.

6. The question is whether science can offer a substitute for the various belief systems upon which social values and structures were traditionally founded. At first it might appear that this is impossible since, again, values being essentially *non objective* cannot logically be derived from objective knowledge. Nor could science propose some 'belief' upon which values might find a new foundation. Science not only ignores, but rejects any assimilation of 'belief' and knowledge. Must we then accept the conclusion that the field of 'true' knowledge and that of values

are and must remain totally separate, absolutely foreign to one another?

7. This profoundly pessimistic conclusion would be erroneous, because in fact there exists between the field of knowledge and the domain of values an inherent historical and logical relationship of *filiation*. Science indeed cannot create, derive or propose values. But the pursuit of objective knowledge is in itself an ethical attitude, founded upon an initial *choice* of a value system, which I shall call 'the ethics of knowledge'. Within this system, the supreme goal, the standard of value, is objective knowledge itself and for its own sake. In deciding to become a scientist, one, explicitly or not, adopts this system and this choice evidently does not logically result from a judgment of knowledge but from a deliberate, axiomatic choice of a standard of value.

8. Historically, the ethics of knowledge, based on the pure postulate that Nature herself is objective, and not 'projective' (as in the Aristotelian system) goes precisely back to the founders of modern science: Galileo, Descartes and Bacon. The construction, by science, of the modern world, and the destruction, again by science, of the foundations of the traditional belief system both ultimately rest upon this initial choice.

9. Hence the moral contradictions which menace modern societies. They have accepted to reap the fruits of science. They have not accepted, hardly have they understood, the ethics of knowledge upon which science is founded. Yet this ethic is the only one able to lay the foundations of a value system wholly compatible with science itself and able to serve humanity in its 'scientific age'.

10. There is no place here to discuss the new humanism which could develop on the basis of the standards of value of the ethic of knowledge. Clearly, however, such a system would serve man in his most unique and precious essence, setting as its primary goal the development of culture and creativity and fighting any form of alienation, whether intellectual, political or economic.

11. I feel that in this moral, intellectual and also political revolution which will have to come if civilization is not to collapse, the scientists have an essential role to play; and that they could not shun this great responsibility without violating their own, freely chosen, moral code.

Discussion

Lal I agree there is no *ultimate* justifying criterion for a value system, but this is also true of scientific statements and it seems to me that this urge for ultimate justifiability is a myth – one doesn't need it. All one wants is the right to be critical.

Monod Well, this is a typical criticism from a typical British empiricist. I think it is just plainly experimentally wrong to say that one does not need an ultimate justification. All social systems have tried to have an ultimate justification or pretended to have one. Of course it was a mythical one. But the fact is that they always tried to have it. Therefore there must be some profound craving for the ultimate justification. And the great difficulty with which we are faced is that if we are truthful as objective scientists, we have to tell people that there is no possibility of ultimate justification of any sort of moral code and therefore we have to accept an axiomatic basis for the social structure and the moral code itself. Let me say that curiously enough there is a parallel here between geometry and ethics. For hundreds or thousands of years it was believed that there was one natural geometry—the Euclidean. I don't think we are as free to choose the axioms for humanity as we are in geometry, but still we must recognize that from the point of view of pure logic – and again I wanted this to be purely a logical argument – we have to recognize, and we have to say, that from now on, if we want to live according to the standards of objective knowledge, we have to agree to say and to teach that the moral code stems from an axiomatic free choice. This is nothing very new that I am saying, in fact this is the attitude of several of the modern existentialist schools, especially of Camus and Sartre.

Levy May I say that the speaker seems to me to be presenting his case upside down. When Professor Monod, draws an analogy between the development of Euclidean geometry and that of ethics and goes on to suggest that we must first seek the appropriate axiomatic basis for ethics and morals he has inverted the order of discovery, mistaken the Greek contribution to mathematical thinking, and obscured the vital distinction between a value judgment and mathematical abstractions. Mathematical propositions like physical propositions are bound together by the same physical logic, and as there is one world so there is one logic common to all peoples. A rational person is one whose thinking fits the world we live in, otherwise there would be no science.

Ethical values and moral judgments are not ideas but feelings evoked directly by social experience. An ethical proposition is concerned with the feelings of a social group, and as there is not yet a unified human society there is no fully fledged international ethic. Ethics is concerned with the discovery of how people have to behave if they are to live together on one earth as one humanity. It is the search for the human counterpart of the laws of nature. Neither can be deduced *a priori*; both require intelligent experience.

Monod Certainly I wouldn't disagree with the fact that experience is important, but I would disagree with going all the way back to a purely utilitarian system. I don't think this will do; I don't think that any system which would not eventually justify somebody sacrificing his own life would be a true ethical system.

Raschid Your central proposition is that science cannot propose values, but the rest of your case is absolutely inconsistent with this. You say that we must arrive at some objective way of getting values but then you evade the whole issue by introducing your notion of 'axioms'. I challenge the dogmatic assertion in the abstract of your talk: 'That while science is not a judge of values in itself, it is very often in a position to give an objective appraisal of the validity of the sources upon which most traditional value systems have sought their justification.' Moral codes can be based on all sorts of grounds – e.g. on humanist considerations, on a Marxist system of values (*Monod*: And see where it leads), on religious premises, on a wide range of philosophical positions and so on. I don't see that science has affected any of these things.

Monod I am not saying that these systems are wrong but that they are outside science and incompatible with it.

Gunnell I should like to ask you to comment on the

pledge to give information,* and in particular your suggestion that a situation might occur in which a scientist should withhold or suppress the results of his research if the implications of that work are dangerous or contentious. Support for thinking that the pledge is valid comes from the number of instances of multiple discovery in science; it is implicit in the nature of science that it will grow and that knowledge discovered and withheld by one worker is going to be separately discovered by another. Therefore, even if he has discovered something which he can see having harmful political consequences, the socially responsible scientist will, nevertheless, feel bound to inform people of what he has discovered in the hope that he will then be able to influence its consequences.

Monod I agree. Nevertheless I think the example of the Einstein–Szilard dilemma is a good one because a certain number of difficulties and problems arise that do not arise at other times. It is still true that in the case of Szilard he was, I think, right in trying to prevent publication of these papers, particularly some by Joliot. And the question also is whether now, thirty years later, we may or not judge the Einstein and Szilard decision to start what in fact was to lead to the Hiroshima bomb. The reason they did decide to go ahead was that they had some information of a similar effort having been started in Germany. I discussed this with Szilard and it is clear that their reasoning was ethical. They were not led by purely nationalist passion – that's one reason why the experiment is so fascinating. They were led to their decision purely by the conviction that were the war won by the Nazis this would be perhaps an almost definitive setback to the progress of culture over the world. Maybe they were right in believing that. Of course they were wrong in their information – in fact the Germans were nowhere and got nowhere – but that has nothing to do with the problem itself. We may find ourselves one of these days in a situation like that.

* 'As socially responsible scientists we hereby pledge not to conceal from the public any information about the general nature of our research and about the dangerous uses to which it might be put. This pledge was accepted at a teach-in organized by members of BSSRS at the 1970 Annual Meeting of the British Association for the Advancement of Science in Durham. It was not accepted by the BSSRS 1970 Annual General Meeting.

Wilkins Speaking as one of the scientists who worked on the development of the bomb, I think it is a general attitude amongst them today that, in view of the very terrible nature of the Nazi régime, they were correct in this policy.

Monod I think we have to realize that when it comes to the really crucial problems then no pledge is going to count – you have to re-examine the whole problem.

Buican Professor Monod, is genetic engineering on human personality justifiable from the point of view of the ethical system you have put forward?

Monod I would say it might conceivably be if we came to a point where we could do something to personality, say to intelligence, in some objectively and rationally measurable and well determined way. However, I think such possibilities are so far in the future that we do not need to worry too much about it. For the time being at least, there are more pressing questions.

Garrood Professor Monod has stated that we must consider the present problems within the present structure of our society. Surely, it is precisely that scientists have failed to alter or improve social institutions and social thinking that creates some of these problems. Science is now beginning to grasp tools that can have very great social implications, and unless scientists can mould institutions they will snap.

Monod I think you do too much honour to the scientists to think that it is up to them to mould institutions. It is up to them to say what is the state of the art in their field, and also to state political and ethical feelings as they are influenced by their profession, but I don't see that they could go much further than that and pretend to substitute for the people, for the lawyers, for the sociologists, for the honest politicians, and so on.

I fully agree that scientists have not been up to the powers that their discoveries have put at the disposal of mankind. This means that science is much more powerful than individual scientists, or even a group of scientists, or even all the scientists that exist at the present time in the world. On the other hand, during the past two hundred and fifty or three hundred years, I think that in the Western world the development of science and the development of democratic institutions have had a great deal to do with one another. The fact that a point of view developed, mostly in science, but it was also infused into society, that no creed is absolute, that no system can pretend to have the whole truth, that any theory should be revised by experiment and observation, has something to do with the democratic principle which recognizes that a democratic society is not perfect, does make experiments, and tries to do better if it can. The two are closely allied and I would not say that any individual scientist had anything very special to do with that, but the general trend of the ideas of science has had a profound influence on the evolution of social institutions and on our attitudes towards them.

Dauman With regard to your point about raising the level of intelligence, it seems to me that if there is any chance that the social problems caused by this discovery would be greater than the benefits, any chance at all, then we should say 'stop'. (*Monod*: I agree.) What is worrying me is that while scientists all over the world are talking about social responsibility, there is hardly anybody who has dared to say 'stop' – they just just haven't got the guts to say 'stop'.

Gibson We are taught, and go on to teach, the omnipotence of research for its own sake and *therefore* have no mechanism by which any research, however abhorrent to the vast majority of the scientific community, can be stopped. Because of our teaching, such a mechanism will be extremely difficult to devise and I believe that both at this conference and elsewhere the BSSRS should give this some serious attention.

Monod I must say that I cannot accept the idea of forcibly stopping research in a legitimate field, where there is knowledge to be gained, and I think that by trying to do that we are not really facing our responsibilities but trying to evade them.

Hayes In connection with this question of science and power, I think, speaking as a biologist, we have power if we group together. I was in West Germany recently and a friend of mine had just started a German genetical society – this, I think, is the only one in West Germany – and there were two points of view about this. One group, including my friend, wanted the society oriented so that questions on the social and political significance of what geneticists in West Germany were doing would be discussed. The other group simply wanted the society to be a rather pleasant institution where they would come and read their papers and have dinners and make speeches. My friend was very worried because at a crucial committee meeting he thought that his group was probably going to lose. In point of fact he won, and the interesting thing is the people he was afraid would vote against him were those geneticists employed by the big and very powerful pharmaceutical and commercial firms in West Germany; in the event they were the very people who supported him because they were isolated individuals, subject to great pressures, economic and otherwise. They felt that the only way they could get help was by support from all their professional colleagues in their country. In this way, if they were confronted with issues, they could at least say: 'It is not really me, I can't help it, but the entire pressure of all my colleagues is the other way.' Therefore, I think that by banding together and forming some communal doctrine with regard to political and social issues, we have power.

Three Fragmentation in science and in society*

David Bohm
Professor of Theoretical Physics
Birkbeck College, University of London

In recent years, there has been a great deal of criticism of science on the part of non-scientists. The articles in the issue of *Impact of Science on Society* having the theme 'Non-scientists Dissect Science' present a spectrum of such criticisms.[1]

The essential point raised in these articles is that science now tends to be characterized by arrogance, dominance, irresponsibility, and by an excessive concentration on esoteric questions, almost equivalent to that of medieval scholasticism. Science, it is claimed, has become something like a new mystery religion, with a priesthood and an occult doctrine that is understood by only a small and select group. It is therefore suggested that there is now a need for general social control of science, to ensure that it will really benefit mankind and to help decrease the gap between science and other aspects of life.

As a scientist, I might be expected to reply to these criticisms. However, I think that such a reply would not be relevant, because the problem is not one that can be restricted to science. Actually, all these features of current scientific activity are manifestations of a general social condition: *fragmentation*. This fragmentation shows itself in nation arrayed against nation, race against race, religion against religion, group against group, and man against man. And in turn, each man is fragmented into different and incompatible loyalties, aims, desires, etc.

In this context, science is partaking of the general conditions of fragmentation. Thus, there are sharp divisions between applied science and pure science, between theory and experiment, between one specialized field and another, and between different branches of each speciality. The gap between science and other

* Reprinted with modifications and an addendum, from *Impact of Science on Society*, vol. 20, no. 2, 1970, p. 159, by permission of Unesco.

aspects of life is just a further example of such fragmentation.

The individual scientist himself is then pulled between all these fragments. In addition, his motivation is fragmented in other more subtle ways, e.g. between the wish to know and understand and the wish to get ahead, to become known, successful, dominant, and to have opinions that are acceptable to his colleagues.

Most of the difficulties that non-scientists are complaining about are essentially of this nature. It would therefore be of no use to set up a social control of science, because the society that would control is beset by essentially the same fragmentation as that which it is trying to control.

Something very deep and pervasive is involved here, something that cannot be discussed adequately in terms of control or organized planning. For such fragmentation is built into the very language of science (which is of course ultimately the outcome of the general language used in society as a whole).

Fragmentation in biology

Consider, for example, the science of biology. A major part of recent research has centred on the analysis of living beings into cells, particles of DNA, RNA, etc. In psychology, each individual human being has often been treated almost like a separate atom, with society regarded as a collection of interacting atoms. And in turn, the individual human being is further analysed as an interacting aggregate of organs, muscles, neurons, control circuits, etc. Thus, our very way of talking and thinking is introducing fragmentation into these fields. It is interesting to compare this with fragmentation in physics which is actually moving in a different direction, away from the notion that the world can be analysed in parts. (See Bohm in *Impact of Science on Society*, vol. 20, no. 2.)

Fragmentation in technology

Fragmentation in science goes together with fragmentation in technology. Each industrial or technical project tends to be regarded as separate. There has been a kind of faith that all these separate actions will somehow work together harmoniously towards general human happiness. But in modern times, this faith has been severely shaken, especially by problems arising in man's over-all participation in nature and in society.

Technology carried out as an accumulation of fragmentary and separate activities had led to a series of crises, including pollution of the natural environment, disturbance of the ecological balance, possible changes of climate, and other serious consequences of a similar nature. Medicine aimed at improving public health and longevity has led to overpopulation and to struggles between peoples and between nations for food and other resources. More generally, different groups that depend on each other technologically have not been able to co-operate. Such co-operation is blocked, because these groups are in different and highly disparate stages of development, with different cultures, different traditions, languages and ways of thinking. Thus, people cannot get together to meet common problems, such as the extreme shortage of food in some areas while there is a glut in others.

Moreover, psychological dangers inherent in overcrowding have recently been pointed out. In connection with these dangers, even technically advanced nations do not seem to be able to deal with the problem of adequate housing, schooling, recreation, etc. Indeed, just these nations show an extreme form of fragmentation, as indicated, for example, by the fact that advanced technology may be carried out in one part of a city, while in another part, there may be a veritable jungle in which it is no longer safe for people to walk in the streets.

Wholeness

From very early times, man has been aware of this general problem of fragmentation and has sought somehow to become whole. For example, consider the root of the word 'health', which is the Anglo-Saxon 'hale', meaning 'whole'. So 'to be healed' is 'to be made whole'. Similarly, the word 'holy' has the same Anglo-Saxon root (related to the German 'heil' and 'heilig'). Thus, man has always wanted to be made whole, not only physically and mentally, but also spiritually. In the effort to become whole, he set up medicine and organized religion, and thus established what may be called 'departments of wholeness', which are, of course, yet another form of fragmentation.

Likewise, science sought originally to give man a wholeness of knowledge and understanding, but ended up fragmented into countless separate fields each of which can overwhelm the student in a flood of specialized information. And new branches of science aimed at bridging the gap between such specialized fields generally end up by adding yet another separate fragment.

The problem of fragmentation is extraordinarily subtle. How can we take steps to deal with it and not just add more fragmentation (as has happened again and again for thousands of years)?

We need to look deeply for very pervasive factors of which we tend to be unaware (as fish may be of the water in which they swim). One of these is evidently our general mode of communication and thinking. As we try to discuss our difficulties, may it not be that our very way of thinking and using language is adding to our problems, rather than helping to solve them?

Linear progress and cycles

Consider, for example, the word 'progress'. Its root means 'to step forward'. The whole of man's technological thinking implies this notion, i.e. to begin in a certain place and to step forward towards a place or a goal which we see as being ahead of us. But when the goal is reached, the problem tends to be dropped. It is generally taken to be someone else's business to go on from there. As indicated earlier, we fall back on a kind of faith that all will somehow work together harmoniously. But, of course, it hasn't. What actually happens is that when we 'progress' from point A to point B, actually a kind of cycle is set up. Things don't just rest there but keep on moving and tend to return to A. For example, if farmers, with a certain goal in mind, alter the species of plants and animals in a certain region, the 'side-effects' of such action are such as will generally further alter the basis on which the steps towards the goal were planned. Thus, clearing land of wild growth to plant more acreage to alfalfa will reduce the population of wild bees needed to fertilize the alfalfa, cutting down yield. The whole of ecology demonstrates the relevance of such cycles in nature.

Similar situations arise in society. Thus, many people are now discussing the problem of violence (which is evidently a manifestation of general social fragmentation). But such discussions generally ignore that violence is a cycle. For example, a given group in society may have grievances, and may act towards settling them by means of a public protest. If their demands are met, they are encouraged to protest more, until the 'other side' becomes frightened and adopts repressive measures. The first group then increases the violence of the protest, which in turn stirs up the violence of the protest, which in turn stirs up the violence of those who 'want to maintain law and order' in the by now almost classical cycle of 'escalation'.

What one can see in such a situation is that any move aimed at some specified goal is fragmentary, and therefore more than likely to activate the cycle, rather than to bring it to an end. Even when particular groups make an over-all plan to meet the problem of violence, they are blocked by their limited ability to understand the problems of other people because of their own fixed opinions and beliefs, which set them apart from the very people whom they wish to help.

There is a similar pattern in pure science. The analysis of the world into parts goes hand in hand with a description of the movement of these parts as a 'process'. The root of this word means 'to go forward', which is, in essence, not different from the meaning of the word 'progress'. Thus, in physics, the laws of movement are expressed by saying that, with the passing of time, something (e.g. a particle) *proceeds* in a certain regular way from A to B. A process is in this way taken as fundamental and cycles are considered to be simply a particular form of process that is analysed as a series of steps in which the movement turns back on itself.

Similarly, in biology, life is described as a process. History, too, is frequently regarded as such a process, which can be studied by analysis into suitable 'elements' and by following the movement and development of these 'elements' in their interactions and interrelations.

On the other hand, the ancients tended to describe all (both men and nature) in terms of cycles, regarded as wholes, in which analysis into partial steps was not taken to be fundamentally relevant. For example, it was supposed that celestial matter is perfect, and that the perfection of its nature should be expressed in circular orbits, because the circle was considered to be the most perfect of all geometric figures. When the observational data failed to fit the assumption of perfectly circular orbits, this led to the consideration of a superposition of epicycles, or 'circles within circles'. In the case given, as in many others, e.g. in mythologies, the specification of a whole cycle was taken to be the fundamental mode of description, in which the succession of particular events and processes played a relatively secondary role.

This approach was, however, too static, as the cycles were

Fragmentation in science and in society

considered to be eternal and repetitive. It thus left no room for a genuine evolution and unfolding of new possibilities and potentialities. In this regard, the notions of linear progress and process were useful, in that they helped lead men to an awareness of the eternally changing nature, first of his environment, and eventually of his society and of himself. Yet the attention was focused on unidirectional movement and something was lost: the awareness that the fundamentally cyclical character of all movement is non-unidirectional.

A language for the fundamentally cyclical nature of movement

Can we experiment with the usage of language and see whether we can adequately call attention to the fundamentally cyclical nature of movement, without falling back into the static character of ancient notions on the subject?

To see what such experimentation might mean, consider, for example, a river, which has its source in various springs and flows towards the sea. Describing a river in this way treats it as a 'process' of movement from sources to the place where it is 'absorbed' and merges with 'something much larger'. But evidently this is only a part of what is relevant here. The water does not actually originate in springs. Rather, it reaches these by seeping through the ground on which it has fallen as rain. The rain in turn came from clouds, which were evaporated originally from the sea. To understand the 'river as a whole' one has to think in terms of such a cycle. This cycle is always going on. Thus, one does not begin by thinking of the cycle as carried out in a series of actions over a period of time. Rather, one thinks in an over-all way of the movements of the various forms of water in a cycle *as a whole*, considered all at once.

To emphasize that a new way of thinking is involved here, it is useful to introduce some new words. Thus, to call attention to a totality, one can use the Greek prefix *holo*, meaning 'whole'. And to indicate that such a 'whole' is to be seen as undivided movement, rather than as a static collection of parts or elements, one can begin, not with a noun, but rather, with the verb form, 'to holocyclate'. This form calls attention to an unbroken totality of movement in many directions at once, and not to a series of fragmentary steps from one place to another, in a definite direction. Thus, to give a holocyclate description of the river would mean to indicate a sort of panoramic view including the sea, the air, the wind, the clouds, the rainfall, the seeping of water through the ground, the emergence of water in many springs, the collection of this water into many rivers, and the ultimate flow of these rivers back into the ocean.

It is being suggested here that, in general, we understand each particular aspect of the movement deeply (e.g. a specified river) only in the context of such a totality of cyclation, in which all aspects originate.

Let us now also introduce the verb 'to procedate', which means 'to think or describe in terms of unidirectional movement' (e.g. in a line). Procedation is in general a useful abstraction from holocyclation. For example, we can follow the flow of a particular river from one place to another by looking at an object floating in the river. We can eventually analyse a 'particular cycle' as a series of such procedative steps. However, this analysis is relevant mainly as a means of providing insight into a totality of cyclation and is not to be considered as providing a fundamental description.

As has already been indicated, the verb 'to holocyclate' does not imply static and eternally repeating cycles. Rather a much more dynamic notion is meant here. For example, with the river, the energy of holocyclation eventually comes from the sun. If the sun's rays were to cease, the oceans would freeze and the rivers would dry up. With energy from the sun, holocyclation is brought into movement. And the movement can be indefinitely varied because of the unlimited number of factors of all kinds (mountains, forests, coastlines, etc.) which are involved.

To symbolize the dynamic and varied nature of holocyclation, one may think of expanding and contracting spiral movements (in contrast to a circular movement which in ancient times most characteristically symbolized a closed cycle). This way of symbolizing holocyclation implies that expansion and contraction of cyclate movements are going on together (e.g. some rivers may be rising to flood while others are drying up). It is also implied that each aspect of cyclation takes place in a larger complex of cyclations (e.g. solar energy, planetary motions, etc.) which are a general context or background.

It is possible to re-examine all of our knowledge, to see what happens when holocyclation rather than procedation becomes our basic mode of description. Thus, in physics, cycles occur in a great many places (e.g. in circuit integrals in electrodynamics, in the action variables of quantum theory, and in many other contexts). At present these are analysed in a series of procedative steps that take place over a period of time, in such a way as to close and form a cycle.

Actually, it is possible to change the language of physics so that a vast totality of such cyclations considered all at once becomes the basic form of description, while procedative steps are taken as abstractions from holocyclation. This implies new kinds of laws, which are not 'laws of motion' in the classical sense. Indeed, the time order of procedative movement is no longer taken as basic. Rather, one discusses in terms of the over-all ordering of cyclation.

In this article there is no space to go into such changes of language, with which the author is now experimenting. However, very new ways of describing space, time, matter, etc. are opened up by the new language forms.

In biology, we can similarly regard life as holocyclation, rather than as a process. Likewise, in psychology and sociology, we can consider man's participation in nature and in society as holocyclation, rather than as a procedation, in which each man proceeds – 'takes steps' – in relation to separate objects in nature and in society, while the latter 'take steps' in relation to him. Such a mode of using language continually focuses attention on undivided wholeness, and on the fact that analysis into a procedation of separate parts is at best a useful abstraction.

Holocyclation and perception

Holocyclation is indeed characteristic of all of our experience. For example, consider modern experiments on perception. The essential result of such experiments is to show that movements of the body are necessary for all perception. That is to say, perception consists, not in seeing objects as isolated entities, but rather, in an over-all awareness of the relationship between the individual's outgoing movements and the incoming sensations. These form a cyclation.

When, for example, a person is given an object which he cannot see to explore by touch, it is found[2] that as his hands and fingers move to handle the object in many ways, a certain knowledge of 'the object as a whole' begins to be abstracted.

Such a person is generally almost unaware of the details of the movements. Yet, as a closer study shows, the knowledge of the object can only originate in an ensemble of relationships between outgoing movements of the hand and tactile sensations which accompany these movements.

At first sight it would appear that in optical perception the situation is different, in that one seems to see the object as a separate entity. However, a closer study shows that this is the case only in situations in which the objects are fairly familiar to us, so that their general forms and properties are given by an adaptation of what has been learned in past experiences with similar experiences with similar objects. Experiments[2] show that a person presented with a new kind of situation in visual perception is unable to see correctly unless he moves his body in various tentative ways and observes how the content of his field of vision changes in relationship to his bodily movements. Thus, optical perception is basically similar in this regard to tactile perception: it involves a cyclation of outgoing movements and incoming sensations.

A very striking example of what has been described above is obtained by considering a person who has been blind all his life, and is suddenly enabled to see (e.g. by an operation). When he first gains vision, he does not perceive objects as such, nor does he generally understand what appears in his field of vision. It takes a long time (possibly years) of relating vision to touch and bodily movements before such understanding develops. But, of course, we tend to fail to notice that something similar has happened with all of us in the first few years of life when we explored the world and thus learned how optical sensations are to be related to bodily movements and tactile perception.[3]

It is implicit in what has been said that time is not very relevant in perception. Indeed, as has been brought out in some detail elsewhere[4] what has been perceived in the past is generally necessary to give form and order to what is perceived in the present. Thus, there is no way to say 'exactly when' we see something. Rather, the ensemble from which a given perceived object is abstracted may in general cover an essentially unspecifiable period of time.

All of this can be expressed succinctly by saying that holocyclation is the fundamental description of perception, while analysis into procedative steps of separate parts or aspects is at most a useful abstraction.

Perception, communication and action

Our general mode of acting and living is evidently determined by our mode of perception and communication, As we perceive and talk, so we will think and act and, therefore, so we will be. Thus, to take the procedation of separate objects as basic is a mode of perception and communication that means a thoroughgoing and pervasive fragmentation in our actions and in what we are. But to perceive and communicate in the general mode of holocyclation will mean an unbroken wholeness in all of life.

To deal with the fragmentation of the individual and of society, it is clearly not enough to propose specific measures in a context in which we still go on with fragmentation in almost all that we see and in almost all that we say. Rather, we have to be alert and observant, ready to experiment with language, and to question even our general modes of perception.

For example, as we meet people who are different from us in background, we will understand that it makes no sense to 'tolerate them' or to 'adapt some of their culture to our needs' or otherwise to treat them as separate beings with whom we are in interrelation. Rather, we see that, in our contacts, whatever these people are becomes 'part of us', and whatever we are becomes 'part of them'. So, whether we like it or not, a cyclation is set in movement, which is actually a new kind of 'wholeness' or 'oneness'. But if we talk, think and perceive in terms of separation, this whole will be experienced as fragmentary and contradictory, because our perception is not in harmony with the cyclation inherent in the actual situation.

Similarly, all movement 'between man and man' is quite generally of the nature of holocylation, as is also the movement 'between man and nature'. When we really understand this, the separative and indivisive approach implied in taking analytic descriptions as basic will be seen to be irrelevant and will thus be dropped. Man and nature can then be perceived as a vast complex of cyclation. This will make possible a right kind of action between man and man and between man and nature, an action that is not one-sided and that is relevant to all significant aspects of the cyclation. Fragmentation will then no longer take place.

When man's action is free of fragmentation, then the general social disharmony described in this paper will tend to die away, rather than to build up and to propagate. Men will see that they live in holocyclation, and it will thus be evident that there is no meaning to efforts to dominate others, to impose one's views or patterns of action on them, or to concentrate on narrow esoteric questions that are of interest only to small and select groups.

Scientists who deeply understand that 'their very being' originates in cyclation with the community and with nature as a whole will, like all others, have the benefit of mankind close to their hearts. Thus, external controls (which are in any case generally very arbitrary) will not be needed. But it is essential to understand that it makes no sense to try to produce harmony in science while society and the individual remain generally in a condition of chaos. Rather, harmony in science has to go together with over-all harmony, and this requires that man shall cease to perceive in a mode of general fragmentation.

Addendum on fragmentation in communication

It can also be said that holocyclation is a proper description of what generally happens in communication. Of course, there is a limited kind of communication that consists simply of conveying information from one person to another, and this can be considered as a linear process. But more generally, communication involves *dialogue*; i.e. a two-way flow.

In such a dialogue, when one person says something, the other person does not as a rule respond with exactly the same meaning as that seen by the first person. Rather, the meanings are only *similar* and not identical. Thus, when the second person replies, the first person sees a *difference* between what he meant to say and what the other person understood. On considering this difference, he may then be able to see something new, which is relevant both to his own views and to those of the other person. So it can go back and forth, with the continual emergence of a new content that is common to both participants. Thus, in a dialogue, each person does not attempt mainly to convey to the other certain ideas that are already known to him. Rather it may be said that the two people are participating in a single cyclical movement in which they are *creating something new together*.

But, of course, such creative communication can take place properly only if people are able freely to listen to each other, without prejudice and without trying to influence each other. Each has to be interested primarily in truth and coherence, so that he is ready to drop his old ideas and intentions, and is able to go into something different, when this is called for. However, if two people merely try to convey certain ideas or points of view to each other, as if these were to be regarded as fixed and final knowledge of fact or reality, then they must inevitably fail to meet. For each will hear the other through the screen of his own thoughts, and will tend to maintain and defend them, regardless of whether or not they are true or coherent. The result will, of course, just be the sort of confusion that leads to the insoluble 'failure of communication' that is now being encountered in almost every phase of life.

Evidently, communication in the sense described above is necessary, both in science and in society as a whole. Indeed, if people are to co-operate (i.e. literally, to 'work together') they have to be able to create something in common, something that takes shape in their mutual actions and discussions, rather than something that is imposed on one person or group by another, by means of superior force or by means of authority.

One of the most important reasons for failure of a true dialogue to arise between people is that each person tends to regard himself as existing separately from the others and from the world as a whole. This way of looking at things is most strongly exemplified by the notion of the mind–matter duality. That is, mind and matter are considered to be separate substances, and different people are supposed to have separate and distinct minds, related by the common material world which provides some sort of mediation between them. Communication is thus thought of as a means of connecting all these different minds, in a process of interaction.

In modern psychology, this point of view is most characteristically exemplified by the meaning of the word *psycho-somatic*. This is derived from the Greek words, *'psyche'*, signifying 'mind' and *'soma'* signifying 'body', or more broadly, 'something that can be touched' (so that it can be taken in this context to signify matter in general). Clearly, the current usage of this word indicates the prevailing analytical point of view, in which mind and body are taken to be separate entities, in interaction. Such a view must inevitably lead to fragmentation in every aspect of life, because it implies fragmentation in each action of the individual human being, whether in relation to nature or in relation to other human beings.

It may be helpful here to suggest a new word for use in this context, i.e. *soma-significant*. The key root in this word is *sign*.

Thus, consider a sign. Because it is made of some material, it is *somatic*. And because it points or indicates beyond itself, it is *significant*. Similarly, a table is somatic, because of its material constitution. And it is significant, as a table, as wood, as coloured, as shaped, as structured in terms of grains, cells, atoms, etc. Evidently, there is an unlimited range and variety of significances, by which we *know* the table. Indeed, any kind of knowledge is ultimately a totality of significance.

It is possible to go further, to point out that feeling, thought, and other aspects of what is commonly called 'mind' are also soma-significant. Evidently, these are significant in that they indicate or point beyond themselves. And as modern scientific investigations show, they are somatic, in the sense that such mental movements are inseparable from physical, chemical and mechanical movements, taking place in the brain, and nervous system as a whole, and in the muscles.

We are in this way led to consider the notion that *all is soma-significant*. That is to say, whatever feature may appear in the context of our discourse or of our thought, it always appears simultaneously, in two aspects, that of soma, and that of significance. The significance points to a broader context, which in turn reflects back in the original context to form a cyclation. These cyclations lead on and on to broader cyclations without limit, and thus ultimately towards the whole content of our discourse and of our thought.

How is fragmentation to come to an end? There is no way at present to give a definite answer to this question. Rather, the very raising of the question is in such a context as to indicate the need for a serious and sustained inquiry. Since fragmentation deeply involves our modes of thinking and using language it seems reasonable, as suggested in the article, to *experiment* with these. Thus, what has been said with regard to holocyclation and soma-significance is not to be regarded as a proposal for some new kind of static metaphysics. Rather, it is a provisional suggestion that is to be considered (like this conference itself) as an aspect of such an experiment. To make this suggestion, and to state it in the form of a talk is an *outgoing action*. To observe what happens and to listen to the response is to be sensitive to the related *incoming action*. If those who hear this talk respond in a similar way, a dialogue will have begun. In such a dialogue, there has to be a harmony between what is actually going on and the significance of what is being said. A similar harmony is needed in broader aspects of the cyclation of soma-significance and of all our actions in every phase of life. To create such harmony is to end fragmentation.

References

1 *Impact of Science on Society*, vol. 20, no. 2, 1970.
2 A summary of these experiments is given in Bohm, D. (1965). *The Special Theory of Relativity*. New York, Benjamin (appendix). See also Gibson, J. G. (1962). *Psych. Review 69*, 477.
3 Piaget, J. (1953). *The Origin of Intelligence in the child*. London, Routledge & Kegan Paul; Bohm, *op. cit.*; Gibson, *op. cit.*
4 Bohm, *op. cit.*; Gibson, J. G. (1960). *Amer. Psychologist 15*, 694.

Discussion

Monod In the 18th century the *Gazette* in Paris gave news such as: 'The philosophers last week have been measuring the temperature of the River Seine with new thermometers.' Philosophy meant science and science meant philosophy up to the beginning of the 19th century. Of the various fragmentations that you mentioned I am convinced that this division between what is called science and what is called philosophy is one of the most dangerous at the present time. On the continent, largely under the influence of German idealists and metaphysics, there has been built a school of philosophy which totally ignores anything scientific in the broadest sense – any sort of objective knowledge. Its practitioners are preoccupied with building imaginary universes about which they can talk in peace without having science intruding. Since knowledge about the real universe has become the field of science, philosophy has to retreat into imaginary universes. In this country and the United States philosophy has retreated in a large part to a sort of ancillary attitude towards science, doing straight epistemology and so on, and rather being afraid of going into some of the most acute philosophical problems. Again this reflects fragmentation in the relationship of man with the universe. One of the most urgent duties of scientists and of philosophers is to contribute to a reunification of their two fields.

Gorinsky I should like to take up the question of unbroken wholeness. I should like to begin with Professor Monod's thought and suggest that an absolute value system can be based on the integrity of nature because if we violate our biological heritage we violate ourselves. We as a species have inherited an equilibrium. We, as scientists, have caused this conflict with the natural environment. So on this I feel that whether we like it or not we have produced this fragmentation, this alienation between man and his biological responsibilities. Unless scientists can safeguard this biological heritage I feel that anything else that they do will be suspect, because after all the ultimate value is survival, both as a species and in terms of a society. At the moment there are other human societies who are termed 'primitive', and who in the name of progress are being destroyed in the Amazon jungles.* They are as alien to us as another species. My point is that the emphasis should not be on recognizing some wholeness which embraces them and us, but on co-existing with these other societies.

Bohm I wouldn't say that. I think our society is not very different from any other which divides and destroys things and that an essential aspect of the question of wholeness is whether or not we are creating a division between ourselves and those whom we feel are the destroyers (who doubtless look on us in the same way). It is absolutely necessary that we should all of us whether 'primitive' or 'advanced' be able to engage in this dialogue of communication that I was talking about so that we cease to break everything up into fragments. If you call somebody else the destroyer you have already introduced fragmentation in your own mind because you are treating his image as the source of evil and so on, and more or less attributing all the evil to somebody else and not taking into account whatever may originate in yourself.

Rothman I want to know on what basis you make wholeness equivalent to harmony. I want to know why you throw away the possibility that one can have unity and within that unity there can be contradictions. If wholeness is harmony, I want to know how change ever occurs. If one is a biologist one accepts the wholeness of the biosphere, one accepts the

* 'Primitive tribes of South America', *Encyclopaedia Britannica Year Book*, 1970, pp. 99–101.

interaction between living organisms, but one is also faced with the fact that this is not a harmonious situation, that species come into being and species disappear, that evolution occurs. What I feel is a weakness in your cyclation concept, as opposed to, say, a dialectical concept, is that it in no way seems to deal with the problem of contradiction.

Bohm I don't see myself how the dialectical analysis deals with it either. Whenever people are in contradiction, whether they believe in dialectics or not, they are always trying to remove the contradiction. If I engaged in a contradictory argument with you you would not accept it. If I said 2 equals 5 you would say nonsense. Perhaps I would say that it is a creative contradiction. Then you have to say no, that's not a creative contradiction, that's one of the other kind that isn't creative. There has been a long discussion around the question: is there a contradiction in a socialist society – they say yes, there is contradiction but it's the creative kind, whereas in the capitalist society it's the destructive kind. But then that becomes very arbitrary because every time you have a contradiction you call it a creative one and when someone else has one you call it a destructive one. Then we ask how are we going to decide which contradictions are creative and which destructive – and actually there's no way to decide. Contradiction is something that happens in our understanding, in our thinking. When we say move south and move north at the same time, that is a contradiction. Fragmentation arises from just that kind of contradiction, namely we are trying to move south and north at the same time. We are giving contrary instructions without noticing it and the result is a confusion and we see tremendous confusion going on all around us, when people say one thing and do another.

Schama I wonder if we might consider the possibility for synthesizing the findings of science in the West with the philosophies of the East. Fragmentation is something which is entirely opposed to the philosophy of the East; everything there, which goes back three to five thousand years, is based on wholeness and I think we have an enormous amount to learn. I think that scientists might well spend some of their time away from their science studying such philosophy.

Bohm It may be true that the philosophy of the East emphasizes wholeness rather than fragmentation. But I think that in another sense the East suffers from one form of fragmentation while the West suffers from another form. If you look at, say, India, it is really very highly fragmented and in many ways the philosophy of wholeness does not lead people to co-operate or work together; on the contrary, they find it very hard to work together. It is pertinent to ask why this has come about. That is: 'Why are these people paying so little attention to their philosophy?' Well, this is a very big question and I don't think we have time to go into it. But I should like to emphasize just one essential difference between the Eastern and Western points of view; it concerns the concept of measure. In the West measure is an essential concept. If we go back to the ancient Greeks, from whom we derive a great deal of our philosophy, measure was in many ways the essence of everything. If we go to the East, measure is almost nothing; the very word '*maya*', which is based on the word 'measure' means illusion; and thus you can understand why there was not much impetus to develop science and technology there. Now, although I think that measure is not illusion, I do think the West over-emphasizes measure. There has been a split, which perhaps happened long ago in history, where on the one side measure was over-emphasized and on the other side the immeasurable was over-emphasized. But to make this division between measure and the immeasurable is another form of fragmentation. That is where I think the trouble arises in the East.

Part two Molecular genetics

Four Molecular genetics: An introductory background

W. Hayes
Professor of Molecular Genetics and Honorary Director
Medical Research Council Molecular Genetics Unit,
Department of Molecular Biology, University of Edinburgh

In this introductory article I am confronted with a formidable task. I have been asked to present, in a very short compass of time, an intelligible picture of the present state of molecular genetics to form a comprehensible background for what is to follow, to readers whose knowledge, I must assume, ranges from virtually nothing to that of those distinguished scientists who have made such outstanding and key contributions to the subject. My predicament may be summed up by reminding you of Bernard Shaw's apology for writing an unduly long letter on the ground that he hadn't the time to write a short one. However, it goes without saying that I will have to be very dogmatic and refrain from giving you the evidence for most of the molecular models I will talk about. You will have to take the detailed evidence on trust. The real evidence, from a pragmatic point of view, is that the models work and have a high predictive value which, after all, is the main criterion of scientific truth – or, perhaps I should say, acceptability. We no longer need to know the conceptual bases for nuclear fusion reactions to believe in their possibility; the fact of the hydrogen bomb is enough.

The structural unit of all living matter is the cell (Figure 1) which consists essentially of a nucleus containing the genetic material in the form of chromosomes; the cytoplasm which contains numerous small bodies called ribosomes and which supports most of the biochemical machinery and activity for building up new cell substance and creating the chemical energy needed for this; and an enveloping membrane through which nutrient materials enter the cell from the environment, and waste and other products leave. This membrane actually turns out to be very complex and to play a key part in co-ordinating cellular activities. Finally, in plant cells and bacteria there is an outer *wall* which protects the cell and gives it its shape.

Our bodies are made up of countless billions of cells such as these; there are estimated to be 10,000 million in the brain alone. All of these are descended from one common ancestor cell, the fertilized ovum. During embryonic development the descendants of this cell, instead of just multiplying, begin to specialize and to become organized into co-operative groups such as skin, muscle, liver and nerve cells, which differ markedly in function and appearance. Speaking very generally, the process of cell specialization is called *differentiation*, while the anatomical organization

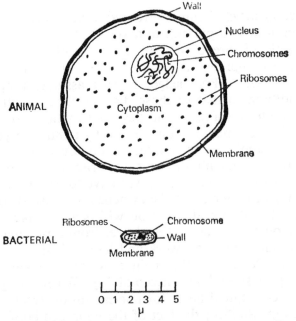

Figure 1 Diagrammatic representation of cells. The wall occurs in plant cells and bacteria but not in animal cells.

of cells into structures and organs is termed *development*. These two processes are found even among such humble multicellular organisms as fungi. However, there are other organisms, such as amoebae and bacteria, whose main attribute is to multiply without differentiation. Since a single bacterium placed in a thimbleful of rich broth can multiply overnight to produce a population of several thousand millions, it is likely that this planet supports more bacteria than the sum of all the cells of all the animals and plants living on it.

But, like animals and plants, there is a great variety of different bacterial species, each with its distinctive size, shape, nutritional needs and other characteristics, and each preserves these distinctions more or less unchanged through countless generations. Thus the cell must be programmed with a specific set of instructions which not only determines the precise way in which each cell performs its function, but is also duplicated and passed on to the two daughter cells at each cell division; that is, the set of functional instructions – the blueprint for the cell – is inheritable, and one of the landmarks in biology was the discovery that the structures responsible for carrying this genetic information and transmitting it to progeny cells are the chromosomes in the nucleus. Thus the genetic material of the chromosomes is the basic biological substance, and the science of genetics, which deals with the structure and behaviour of the genetic material, is the basic biological science.

Now classical genetics, which originated with the work of Mendel just over a hundred years ago, together with miscroscopic observations of the behaviour of the chromosomes, has been mainly concerned with elucidating the *rules* of inheritance and applying them to the important problems of animal and plant breeding. It turned out that the basic determinants of characters, the genes, are arranged linearly on the chromosomes, and that the expression of the characters may be lost or modified by alterations in genes called *mutations*. The basic method of genetic analysis, put in a very simplified way, is to obtain mutant organisms which are altered in two or more characters, say a and b instead of A and B (Figure 2), to cross these (by means of sexual matings) with unaltered or wild-type strains, and to examine the distribution of the character differences among the progeny of the cross (Figure 2). As you will see from the left-hand side of the diagram, if the determinants of two characters (such as A and C) are located on different chromosomes (Figure 2), the character differences will be inherited independently and randomly among the progeny, because the chromosomes are so distributed; but if the two characters (such as A and B) are located on the same chromosome, the two characters typifying each parent will tend to be inherited together, i.e. A with B and a with b.

However, there is another process known as crossing-over or *recombination*, shown on the right of Figure 2, whereby parental combinations of genes *on the same chromosome* can be exchanged

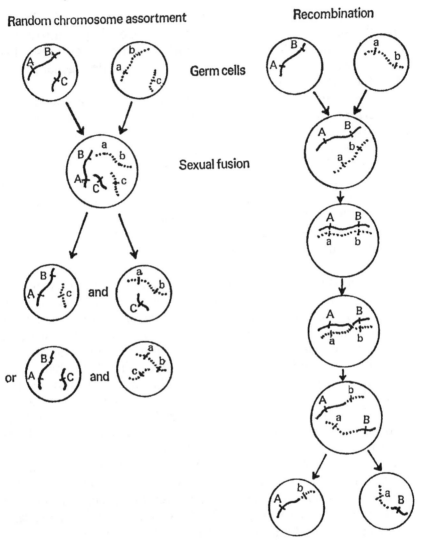

*Figure 2 Left: random chromosome assortment;
right: recombination during the production of new germ cells.*

during germ cell formation. The two parental chromosomes come together and undergo a break and crosswise reunion, so that the progeny inherit part of the chromosome from one parent and part from the other. Now, if two genes are far apart on the chromosome, an exchange occurring almost anywhere will separate them so that they will tend to be inherited randomly as if they were on separate chromosomes. But the closer they are together the less likely they are to be separated by an exchange and the more frequently the combinations found in one or the other parent will be inherited together. Thus we can assess the relative distances between two genes on the same chromosome by counting the frequency with which parental combinations of characters (A and B, or a and b) occur among the offspring. If two genes are extremely close together it may be necessary to examine many thousands of progeny before a recombinant is found, and this is why animals and plants are unsuitable material for genetic analysis if we want to study the relationship of mutations at high resolution – that is, to study the fine structure of the chromosome on a molecular scale.

Before I leave these basic considerations I want to make two more points. One is this – if we introduce into the same cell two alternatives of the same gene, such as *A* and *a*, we usually find that only one of these, *A*, is expressed and so we say that *A* is dominant to *a*. There is nothing mysterious about this. It simply means that gene *A* is producing its normal product while *a* is prevented from producing anything by a mutation. There are other cases where mutant genes determine the synthesis of *altered* products; a good example is sickle-cell anaemia which results from the production of a faulty haemoglobin as a result of mutation in a haemoglobin gene. In individuals who inherit a normal gene from one parent and a mutant, sickle-cell gene from the other, both types of haemoglobin are found in the red blood cells.[1]

My second point concerns the fact that few characters stem from the expression of a single gene. For instance, even quite simple biological products may emerge from a *series* of biochemical reactions, each determined by a different gene. Thus the same character change in two organisms could be the result of mutations in the same gene or in different genes. How can we distinguish these two possibilities? If the mutations are in different genes we might be able to show this by recombination analysis, but if they are very close together the method may not work. The answer is what is called a *complementation test* in which the two mutant genes are introduced into the same cell by mating the parental organisms which carry them. If the two genes are the same, the organisms will have no good copy of this gene and, therefore, will be mutant. But if the genes are different, one good copy of every gene will be present so that the organism will show the normal instead of the mutant character (Figure 3).

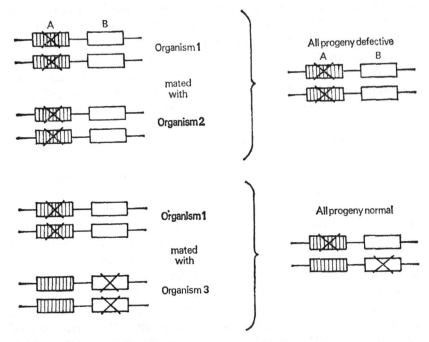

Figure 3 The complementation test. All three organisms are assumed to be homozygous for both genes A and B, i.e., the alternatives for each of these genes are identical. (If the alternatives for a particular gene differ, the organism is said to be heterozygous for that gene.) It is assumed that the expression of a particular characteristic in the organism requires the proper functioning of both genes A and B. By crossing these organisms it is possible to deduce whether the reason they do not express this character is due to a defect in the same gene (e.g. in A in both organisms) or in different genes (i.e. A in one organism and B in the other).

What, then, is *molecular* genetics? The simple answer is that it uses the methods of genetic analysis, which I have just described, to explore the structure and function of the genetic material at the molecular level. As a starting point we must ask, What is the molecular composition of genes? and what is their function in molecular terms? The answer to the second question came first. All living processes are mediated by specific protein catalysts called *enzymes* which act on nutrients entering the cell from the environment, building them up into new structural components, breaking them down to form building blocks, or oxidizing them to obtain energy for these syntheses. Given the necessary nutrients and enzymes, and some kind of co-ordinating system, growth of the cell becomes almost an automatic operation. The real beginning of molecular biology was the demonstration, about thirty years ago in the fungus *Neurospora*, that what genes did was to determine the specific structure of enzymes, formulated in the famous one gene–one enzyme hypothesis of Beadle and Tatum.[2] Thus the formerly separate sciences of genetics and biochemistry came together for the first time through the marriage of their basic functional units, the gene and the enzyme, and all subsequent progress has stemmed from their continued co-operation, although biophysics has recently come to play an increasingly important role in the partnership.

The next major step forward, which brought us to our present state of knowledge, also resulted from parallel developments in genetics and physico-chemical analysis. First of all, bacteria were found to possess properties which are ideally suited to the analysis of genetic function at the molecular level: (1) They have generation times which may be as short as twenty minutes; (2) they form large homogeneous populations which are very amenable to biochemical analysis; (3) they are haploid organisms, that is (unlike animal and plant cells) they have only one set of genes, so that recessive mutations are immediately expressed and can be recognized; (4) simple biochemical characters such as ability or inability to synthesize amino acids (the building blocks from which proteins are constructed) or to break down specific sugars, which arise directly from gene action, can be studied; (5) in many species of bacteria primitive sexual systems are found whereby genes can be transferred from one cell to another so that recombination analysis and complementation tests are possible. In such systems methods are available for screening and counting extremely rare recombinant types among millions of other progeny of a cross. For example, if we cross a strain of bacteria which requires substance A but not B for growth, with another strain which requires B but not A, the only progeny bacteria which can grow and produce visible colonies, on a jelly lacking both substances A and B, are those in which the functional B gene from one parent and the A gene from the other have been linked by recombination. By using such methods it is possible to estimate the arrangement of, and relative distances between, mutational sites within single genes at dimensions of molecular size.

It is perhaps appropriate to mention here the nature of the

collaboration between genetic and chemical methods in molecular biology. The approach of the chemist and physical chemist is to break open the cells, to isolate and purify their components (which may be either normal or altered by mutation), and finally to analyse their molecular composition and structure in the hope that this will yield information about their function and behaviour. Genetics, on the other hand, allows one to establish spatial and functional relationships between genetic determinants *as they actually operate in the living cell*. The relative importance of these two approaches has varied enormously. For example, the elucidation of the structure of DNA – by far the most important biological discovery of this century – revolutionized our concepts of the nature and function of the genetic material, as we shall see. But it is only fair to say that the spur which drove Wilkins, Watson and Crick to analysis of DNA in the first place was the much earlier observation that this substance, in highly purified form, could mediate genetic transfer between bacteria, and was therefore likely to be the genetic material itself.[3] On the other hand our knowledge of how genetic activity is regulated, which I will talk about shortly, came almost entirely from concepts derived from genetical experiments which were later confirmed by biochemical methods.

I must now move on to review briefly some of the major achievements of these joint studies and will begin with DNA, which is short for deoxyribonucleic acid (Figure 4).[4] A molecule of DNA may be represented diagrammatically as a ladder which is twisted so that the two uprights are wound around one another to form a double helix. Those uprights, which are really flexible threads, have a simple and uniform chemical composition. Attached to the uprights at regular intervals, and facing inwards, are flat nitrogen-containing molecules called bases, of which there are four kinds, called adenine, thymine, cytosine and guanine, and abbreviated A, T, C and G. The bases of each strand are weakly bonded to the bases of the opposite strand to form base-pairs. Thus the base-pairs form the rungs of the ladder, but it is as if each rung was cut and the two halves joined by glue so that they are easily pulled apart. Because the bases differ in size and molecular arrangement, in order to preserve the shape of the structure, there is a strict rule that A can pair only with T, and C only with G (Figure 5). We will see the importance of this shortly.

Now the first property of genetic material is to carry information and since the only irregularity in the DNA double helix is the sequence of bases, or of base-pairs, along the long axis of the molecule, the information must be coded by this sequence (AGTTGAC in Figure 5).[5] I have already mentioned that the function of genes is to determine the specific structure of protein enzymes, so we must look for a clue in the structure of proteins. It happens that these consist of long chains of rather simple chemical subunits called amino acids, of which there are twenty different types. An average protein chain may contain about 300

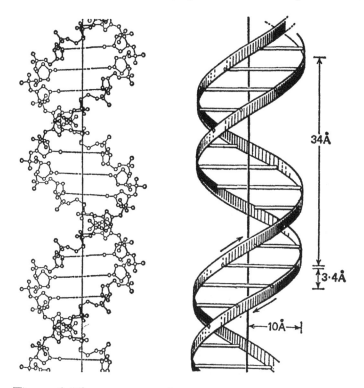

Figure 4 The structure of DNA. Right: schematic diagram in which the base-pairs are represented by steps in the spiral staircase structure and the sugar-phosphate chains which link them together by the banisters; left: as for the figure on the right except that the atoms in the sugar-phosphate chain are shown. (1 cm = 100,000,000 Å)

amino acids, but these chains are not loose like a chain of different shaped beads, but are folded up into rather globular molecules of different configurations. And here we arrive at the heart of the matter, for the specific activity of proteins is a function of their folded configuration and this, in turn, is determined by the specific sequence of amino acids in the chain. It follows logically that the problem of the genetic code is this: How can the sequence of four bases in the DNA specify the sequence of twenty amino acids in a protein? Theoretically this proved a simple problem since four bases can be arranged in sixty-four different triplet sequences, ATC, TAC, CAT, TCA, CTA, ATG and so on. As a result of rather complicated genetical

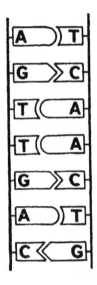

Figure 5 Schematic view of DNA showing specific base-pairing between adenine (*A*) and thymine (*T*) and between cytosine (*C*) and guanine (*G*).

and chemical experiments it turned out that this is indeed the case, and that the sequence of amino acids in a protein is determined by the sequence of triplets of base-pairs in the DNA[6]

Since three base-pairs code for an amino acid and there are about 300 amino acids in the average protein, it follows that a gene determining a protein is about 1,000 base-pairs long. On this basis, the DNA of a very small virus, less than two thousandths of a millimetre long, could determine the synthesis of only three or four proteins and this has been confirmed experimentally. Again the chromosomal DNA of the bacterium *Escherichia coli* forms a circle just over a millimetre in length and could code for about 4,000 proteins. In comparison, the DNA

Molecular genetics: An introductory background 49

in the chromosome of a mouse cell would extend to nearly two metres and have the potentiality to specify four million different proteins!

The second function of genetic material is the ability to reproduce itself or replicate.[7] Replication is a built-in property of the DNA molecule. As you will see from Figure 6, all that is needed is for the two helical threads to unwind and separate. Newly synthesized bases, connected to thread fragments, can then bond specifically to the old bases in accordance with the rule A to T, T to A, C to G and G to C; the thread fragments are then joined together by an enzyme called replicase to form two

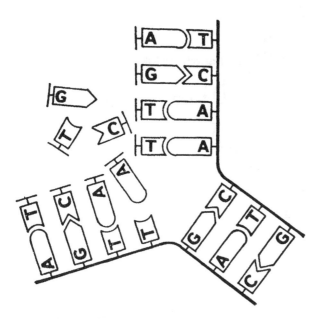

Figure 6 Specific base-pairing as the basis for duplication of the DNA molecule.

daughter double helices, each identical to the original double helix and composed of one of its original threads and the new one. Elegant physical experiments have shown that this mechanism indeed operates and that the whole chromosome of certain bacteria, containing about four million base-pairs, reproduces itself in just 40 minutes. Replication is known to start at a specific point on the DNA of the bacterial chromosome and an interesting corollary of this, illustrated in Figure 7, is that there are twice as many copies of genes near this starting point as of genes near the terminus of replication. Moreover, bacteria given a rich food supply can grow very quickly and divide every 20 minutes – only half the time required for chromosome replication. In order to compensate for this and to ensure that a chromosome is available for every daughter cell a signal, which appears to be a function of the cell volume, triggers off a new round of chromosome replication when the first round is only half completed. This

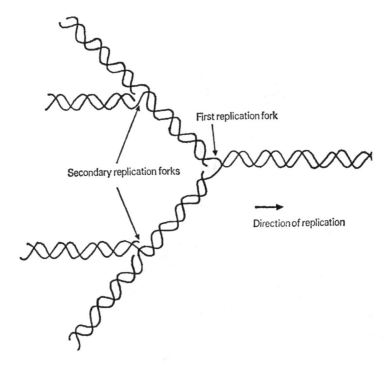

Figure 7 The DNA helix on the extreme right of the figure is being duplicated at the first replication fork to produce two identical daughter molecules. In dichotomous replication, duplication of these daughter molecules is begun before the first replication is finished. The net result is that there are four copies of genes near the extreme left of the DNA molecule, two copies of those between the secondary and first replication forks, and only one of those to the right of the first replication fork.

is called dichotomous replication and leads to the situation shown in Figure 7, in which there are four copies of genes near the starting point, with an equivalent increase in the production of proteins determined by them, for every one at the terminus.

The third and last function of genetic material is that it should be capable of mutation,[8] since mutation is not only a known fact but is also an essential requirement for evolution. Mutation is accounted for by occasional errors during replication which result in the substitution of one base-pair of the DNA by another, thus changing a triplet coding for one amino acid into a different triplet which codes for a different amino acid or, sometimes, for nonsense. In the former case the function of the protein may or may not be altered, depending on the particular amino acid substitution involved; but in the case of a nonsense mutation the protein structure is usually wrecked.

The next question is: How is the information contained in the genes actually translated into protein structure? If we feed cells with radioactively-labelled amino acids and then separate the various cellular components and look to see what components are associated with newly formed radioactive proteins, it turns out that proteins are mainly synthesized, not in the nucleus or in association with the DNA, but on the numerous small ribosomal bodies in the cytoplasm. How, then, does the genetic message reach the ribosomes from the DNA? The messenger was found to be another type of nucleic acid called ribonucleic acid or RNA, which consists of only a single thread and, in bacteria at least, tends to be rather short-lived. Under the influence of an enzyme called RNA polymerase, this *messenger*-RNA is made by copying the base sequence from one of the chromosomal DNA threads into its own base sequence, in much the same way as new DNA threads are synthesized during replication; it then strips off and migrates to the ribosomes (Figure 8). Thus the ribosomes now carry a linear sequence of base triplets which specify the order of the various amino acids in the protein. But how do the various amino acids recognize their specific base triplets on the messenger-RNA so that they can line up in the proper order? This is accomplished by another type of very short RNA molecules known as *transfer*-RNAs. Each amino acid has a species of transfer-RNA of its own to which it binds. At a different site on the RNA molecule is a triplet of bases complementary to the triplet on the messenger-RNA which codes for the amino acid it carries; these pairs of triplets thus recognize each other and come together according to the same general rules which govern the replication of DNA. In this way the amino acids are aligned in the correct order and are finally joined together by an enzyme to form a chain which, on release from the ribosome, folds into its specific protein configuration (Figure 8).

Now the world of bacteria, like our own, is a highly competitive one which involves fierce competition for limited natural resources. We would therefore expect that evolution would have

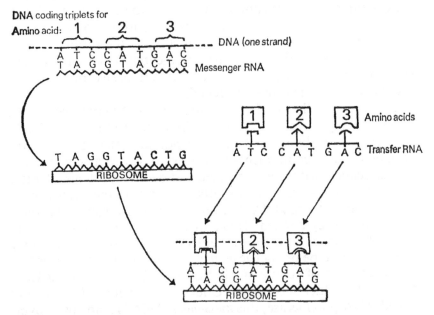

Figure 8 The synthesis of a messenger-RNA with a base-sequence complementary to that of one strand of a DNA gene and its translation at the ribosome into the protein whose amino acid sequence is coded by this gene. The amino acids are distinguished by numbers 1, 2 and 3. The base represented by T in RNA has the same base-pairing properties as that represented by T in DNA, but a slightly different chemical structure.

provided bacteria with refined mechanisms for streamlining their biochemical economy, so that all the genes are not working at full pitch, churning out many proteins which may only be required periodically. But no controlling mechanism can evolve and have any value unless it is inheritable. We can therefore predict that the control of biosynthesis is rooted in the genetic material and, therefore, is subject to disruption by mutation and amenable to genetic analysis. One of the most important control mechanisms elucidated by genetic analysis is the operon model of Jacob and Monod which I must tell you about briefly (Figure 9). In essence, an operon consists of (1) a number of genes, S_1, S_2, S_3 (although there could be just one) which determine the structure of proteins; (2) an adjacent region of DNA called the promoter, P, which is recognized by the enzyme which copies the genes into messenger-RNA so that the genetic information of the genes can be translated into proteins; (3) a regulatory

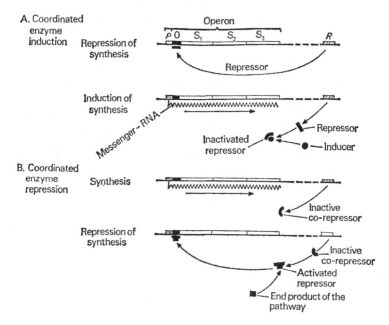

Figure 9 *Mechanisms for controlling the 'read out' of the genetic information.*

gene, R, not necessarily close to the operon, which determines the synthesis of a cytoplasmic *protein repressor* which switches off the activity of the operon by specifically attaching to (4) an operator region, O, which lies between the promoter and the genes so that their transcription into messenger-RNA is blocked.[9]

This model can work in two ways. Firstly, suppose that the functions of genes S_1, S_2 and S_3 is to make enzymes which break down the sugar lactose as a source of energy, so that activity of the genes is purposeful only if lactose is present in the environment. It turns out that in the absence of lactose the genes are switched off by the repressor; but if lactose is present it acts as an *inducer*, specifically combining with the repressor and altering its shape so that it loses its affinity for the operator. The result is that the genes are switched on only when their activity is required. An alternative case is where the enzymes determined by the genes make some molecule needed by the cell, but which may be produced in excess of requirements. In systems of this sort the product of gene *R* is not itself a repressor, but is activated to become a repressor by an excess of the end-product, so that the genes are switched off when they are no longer needed.

Mutations affecting the regulatory gene, *R*, may wreck the repressor protein which can then no longer combine with the operator, so that the genes are switched on all the time. Alternatively, a mutation may alter the repressor protein in such a way that it combines with the operator normally but loses its affinity for the inducer – say, lactose – and in such a case the genes are permanently turned off. Altered repressors of both types have now been isolated and purified. Mutations in the operator have also been described and these result in absence of repression, even if the operon is transferred to another cell whose cytoplasm contains active repressor. On the other hand, mutations in the promoter do not affect the induction or repression of gene activity, but only the rate at which the genes act depending on whether the affinity of the promoter for the messenger-RNA synthesizing enzyme is increased or diminished.

Operon control of gene activity has now been widely reported in many bacteria and viruses, and there is some evidence that it may also occur in animal cells. It has also been invoked as a possible model to explain differentiation since it allows the activity of large blocks of genes, controlling different functions, to be switched on or off together. But perhaps the most important aspect of the model is its clear demonstration that there exist proteins whose affinity for one type of molecule can be activated or repressed in very specific ways by other, unrelated, molecules. Such proteins are called *allosteric* and can serve as biological relay systems whereby the activity of different biochemical pathways can be co-ordinated.[10] In fact, hormone action can best be explained on this basis. The operon system is really a negative once since the key molecule is a repressor. However, positive systems have also been described in animal cells as well as in bacteria and viruses, in which various protein and other factors are required for activation of the enzyme which makes messenger-

RNA, for example, to enable it to recognize specific promoters. One of these factors, called cyclic-AMP, also plays an important role in the activity of hormones such as adrenaline. On the other hand, mutational changes in the promoter can obviate the need for cyclic-AMP.

I may mention here that Shapiro and Beckwith and their colleagues, by clever manipulations of a rather restricted kind, have succeeded in isolating the lactose operon of a bacterium in a pure state,[11] so that it can be studied in the test tube, while Khorana, using a mixture of biological and chemical methods, has managed to synthesize a small gene, about 80 bases long, which determines the structure not of a protein, but of a species of transfer-RNA.[12]

I wish to conclude this talk by saying a few words about viruses, and especially about bacterial viruses.[13] Viruses consist only of genetic material, which is usually DNA but may be RNA instead, wrapped up in a protective protein coat. Virus infection is a genetic phenomenon, the essence of which is the introduction of the viral DNA into the host cell. In the case of most bacterial viruses the protein coat remains outside the cell. The viral DNA carries in its sequence of bases all the information necessary for its own replication and for the synthesis of new viral proteins, and simply uses the energy production mechanism and protein-synthesizing machinery of the cell for these purposes. This cellular machinery is quite neutral and translates the viral information as if it were its own. When all the viral components have been made in adequate concentration they are assembled, by a kind of automatic crystallization, into new virus particles which then escape from the cell, usually by bursting and killing it. In the case of RNA viruses, the viral RNA goes straight to the ribosomes where it acts as its own messenger.

Viruses which invariably kill or damage their host cells in this way are called *virulent*. But among the bacterial viruses perhaps a majority, which are called *temperate*, have a second option open to them. Following infection, the virus DNA produces a protein repressor which usually switches off those operons concerned with its viral functions before they get started. Thus the virus DNA becomes a dormant *provirus* molecule which comes under cellular control, replicating in step with the bacterial chromosome and being inherited by daughter bacteria.

Populations of bacteria which carry a provirus in this way display two properties. First, repressor control breaks down in an occasional cell, usually for physiological and not genetic reasons, so that the virus becomes virulent and bursts the cell with the result that virus particles are released and are always found in the environment. Were it not for this the existence of the provirus could not be recognized. However if this breakdown of repression should happen because a mutation has occurred in the virus repressor gene, then the liberated particles will be virulent viruses; they can no longer produce repressor to shut off their genes when they infect new cells. Secondly, populations of bacteria carrying provirus have repressor already in their cytoplasms and so are immune to infection by any virus which is sensitive to this repressor.

A last question is: How is the provirus carried? In some cases the provirus appears to be replicated by the cell like a small independent chromosome, but we are not yet quite sure what the mechanism is. In other cases the viral DNA, which is a circular structure, becomes inserted by recombination into the continuity of the circular bacterial chromosome as Figure 10 shows. It is thereafter replicated as a part of the chromosome. By a reversal of this process the provirus can emerge again, but here an interesting accident can happen, due to a faulty recognition of just where the recombination event should occur, with the result that part of the bacterial DNA, and the genes it carries, becomes incorporated into the virus DNA and, subsequently, into a virus particle (Figure 10). These bacterial genes thus acquire the viral properties of infectivity and independent replication, so that they are transmissible and many copies of them may be found in infected cells. Again, if a population of bacteria which have congenital defects, due for example to mutations in genes determining the synthesis of essential growth factors, are infected with viruses which carry good copies of these genes, the congenital defects will be remedied in an inheritable way in those cells which insert the virus DNA into their chromosomes.[14]

Genetic entities like the DNA of temperate viruses which can replicate in two ways – either in the cytoplasm under their own control, or as part of the bacterial chromosome – have been called *episomes* or plasmids. There is evidence that some animal viruses, such as certain tumour-producing viruses and herpes simplex virus, behave in the same way. In certain bacterial species there is another class of episome typified by the so-called *sex factors*, which are small circles of DNA with no extra-cellular existence and no protein coats.[15] Acquisition of one of these factors turns a bacterium into a male so that it can now mate with, and transfer the factor to, female bacteria which lack it, so that they too become males. This type of sexuality is, therefore, infectious. When these sex factors become inserted into and are

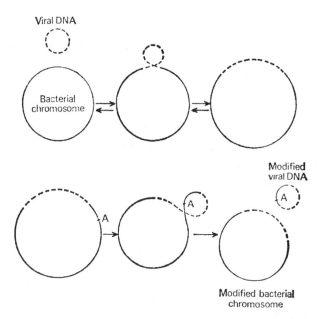

Figure 10 Top: showing how the circular DNA of a temperate virus or plasmid may be inserted into (and subsequently deleted from) the continuity of the circular DNA of the bacterial chromosome by recombination (genetic exchange); bottom: showing how an error in recognition of the proper regions for recombination may generate a virus DNA molecule which carries a part of the bacterial chromosome in it. The interrupted line indicates the virus DNA and the continuous line the bacterial DNA. A represents a bacterial gene.

then released from the chromosome, they too can pick up adjacent bacterial genes which then become part of the sex factor and are transmitted with it to other cells. Sex factors have recently acquired considerable importance in medicine, since many have been found to carry genes conferring resistance to whole groups of antibiotics used in the treatment and prevention of infectious diseases, which they can disseminate widely throughout bacterial populations, making them multiply resistant.

Since these discussions are concerned with the social impact of modern biology, I would like to conclude with a message about the social relevance of what I have said. My message is a simple one: it is that molecular biology has clearly shown that the phenomena of life at the cellular level can be entirely and exclusively accounted for by the known behaviour and interactions of molecules, so that vitalistic ideas, so prevalent until recently, have become irrelevant and unnecessary. I see no reason to believe that this situation will change as the methods of molecular biology penetrate the mysteries of differentiation and development, and of neurobiology and pharmacology, as they are now beginning to do. There can be no doubt that this new vision of ourselves as merely the very complex, and perhaps even predictable, end-product of an exclusively macromolecular evolution, will exert as profound an effect on our social, ethical and political attitudes as have the enlightenments of Darwin and Freud.

References

1 Allison, A. C. (1954). 'Protection afforded by sickle-cell trait against sub-tertian molarial infection', *Brit. med. J.* i, 290; Baglioni, C. and Ingram, V. M. (1961). 'Four adult haemoglobin types in one person', *Nature, Lond. 189*, 465.
2 Beadle, G. W. and Tatum, E. L. (1941). 'Genetic control of biochemical reactions in Neurospora', *Proc. natl Acad. Sci., Wash. 27*, 499.
3 Avery, O. T., Macleod, C. M. and McCarty, M. (1944). 'Studies on the chemical nature of the substance inducing transformation of pneumococcal types. I. Induction of transformation by a deoxyribonucleic acid fraction isolated from pneumococcus Type III', *J. exptl Med. 79*, 137.
4 Watson, J. D. and Crick, F. H. C. (1953). 'The structure of DNA', *Cold Spr. Harb. Symp. quant. Biol. 18*, 123.
5 Watson, J. D. and Crick, F. H. C. (1953). 'Genetic implications of the structure of deoxyribonucleic acid', *Nature, Lond. 171*, 964.
6 Stretton, A. O. W. (1965). 'The genetic code', *Brit. med. Bull. 21*, 229 (review article); Khorana, H. G. (1968). 'Synthesis in the study of nucleic acids', *Biochem. J. 109*, 709.
7 Watson and Crick, *op. cit.* (see n. 4).
8 *Ibid.*
9 Jacob, F. and Monod, J. (1961). 'On the regulation of gene activity', *Cold Spr. Harb. Symp. quant. Biol. 26*, 193.
10 *Ibid.*

11 Shapiro, J., Machattie, L., Eron, L., Ihler, G., Ippen, K. and Beckwith, J. (1969). 'Isolation of pure *lac* operon DNA', *Nature, Lond. 224*, 768.
12 See 'News and Views' (1970). *Nature, Lond. 227*, 13.
13 See Stent, G. (1963). *Molecular Biology of Bacterial Viruses.* San Francisco & London, W. H. Freeman.
14 *Ibid.*
15 See Hayes, W. (1966). 'Genetic Society Mendel Lecture: Sex factors and viruses', *Proc. Roy. Soc., B. 164*, 230.

For an excellent elementary review of this whole field, see J. R. S. Fincham (1965). *Microbial and Molecular Genetics.* London, English Universities Press.

Five Molecular genetics: short-term applications and long-term possibilities

M. R. Pollock
Professor of Biology,
Department of Molecular Biology,
University of Edinburgh

In the preceding article Hayes has described how the genetic constitution ('genome') of a cell, in the form of immensely long threads of DNA composed of millions of so-called bases strung together linearly in a uniquely specific sequence, carries the information for the construction of very large numbers of different proteins, most of them enzymes, each of which catalyses one particular important chemical reaction in the cell's metabolism.

Different coherent DNA stretches ('genes') thus code for different enzymes and, though physically attached to each other in the form of chromosomes and often acting co-ordinately especially during replication, functionally they can be regarded as relatively independent of each other. An enzyme-coding ('structural') gene contains all necessary specific information for the construction of a particular enzyme. In suitable conditions – that is if provided with a proper supply of energy, non-specific building-blocks (such as amino acids) and certain ancillary helper molecules and complexes (such as transfer RNAs, ribosomes[1]) – it will specifically promote the synthesis of its 'own' enzyme. When we think of a gene, then, we generally think of an enzyme: that is, in functional terms, of a particular biochemical reaction, such as the splitting of lactose or one step in the chain of reactions for the oxidation of glucose (to provide energy) or the synthesis of an amino acid (for the formation of proteins).

As well as structural genes, the cell's genome also contains control ('regulatory') genes, rather less well understood, which determine the *rate* at which any particular structural gene, or group of such genes, functions in producing its enzyme. It is the sum of all these chemical (and physical) reactions, stringently controlled and integrated, that make up the individual character of an organism.

Most of what is now known about how genes function and interact at a molecular level has been obtained from work with micro-organisms – mainly bacteria and their viruses. It is therefore not surprising that some of the earliest and most promising work on gene manipulation – attempts, that is, to alter the types and functioning of genes within an organism – has been done with bacteria and bacterial viruses ('bacteriophage').

Most of this article will, therefore, be concerned with applications of molecular genetics in bacteria.

However, the fundamental principles of the subject apply equally to higher organisms and during recent years research workers have naturally been considering possible applications to medicine.[2] It will, therefore, be appropriate to consider what possibilities may develop in that direction in the immediate future and try to present a balanced assessment of the prospects.

With lower organisms, objectives in genetic manipulative research are, or should be, concentrated on harnessing their potentialities for the benefit of human beings, whereas in man they are, or ought to be, directed towards the repair of dangerous, genetically determined defects. The objectives are, therefore, very different in the two cases; but the theoretical foundations and, to a large extent, the technical procedures employed are the same.

The idea of *manipulating heredity*, particularly as applied to man and when phrases such as 'genetic engineering' are used, often engenders a terror amongst non-scientists. This is partly no doubt because of misleading and exaggerated suggestions put forward by inexpert writers in the popular press as well as irresponsible and premature conclusions made by scientists themselves.

Partly, too, this fear is engendered by the self-propagating character of changes initiated in the genetic system: anxiety that something newly initiated may go on for ever and perhaps even get out of control like an epidemic and destroy or permanently warp the human race. However, it should be appreciated that genetic manipulation in the more traditional sense of animal and plant breeding to construct economically superior varieties of stock for food production is fully accepted as ethically permissible and, indeed, of vital importance to humanity. The recent 'green revolution' in the tropics which has increased crop yields so strikingly during the last ten years derives from the development of new plant varieties by the use of genetical techniques that, in a sense, could be regarded as intermediate between classical and molecular.

One thing should be said here. I do not believe there is, has been, or could ever be, a scientific discovery potentially of benefit to humanity that might not be exploited to some degree, *against* mankind, if people want to do so. Similarly there is no planned research project judged to be worth while as possibly beneficial to the community, that could not be turned against society, if desired. There may be times when – with shortage of resources – certain types of research ultimately of value, should be discouraged in favour of more urgent or more opportune projects. But it is the direction in which applications of fundamental scientific discoveries are turned, not their nature, which has to be controlled.

Principles of genetic modification

Possible types of alteration

Genetic alterations can be quantitative or qualitative. That is, variations may be in the extent of expression of a particular gene or genes (the rate of formation of an enzyme, for instance) or in the type of gene expressed (where the gene product is modified or additions made to the genetic repertoire).

QUANTITATIVE CHANGES

1. *Derepression.* Many genes are not normally fully switched on and may operate (for reasons of internal balance or economy in the cell) at a fraction (e.g. less than 1 per cent) of their maximum potential. Some genes have switches sensitive to specific factors in the environment which, if present, can turn the switch on ('induction') or off ('repression'). Their effects are transitory, however, and the gene reverts to its former level of expression as soon as, or shortly after, the external factor disappears. These changes are not, of course, genetic; they are not inherited and do not directly concern us here.

However, the same switches as those involved in these phenomena of induction and repression (i.e. so-called 'enzyme adaptation') as well as others, are susceptible to genetically stable alterations which may fix them in the open (or shut) position indefinitely. Often minor alterations to the DNA, a change in a single nucleotide for instance, may be sufficient.

Thus, genes normally running at 'trickle' rate, or those dependent for full expression upon very particular factors in the environment, may be allowed to operate maximally at all times regardless of the medium. In bacteria the yield of certain enzymes may be increased several hundred fold by such genetic *derepression*.

2. *Multiple gene copies*. Another possibility is an increase in the number of copies of a particular gene in each cell. This can occur in several different ways.

Normally, bacteria possess only a single copy of each type of gene per cell (i.e. they are 'haploid'). In higher organisms there are at least two (i.e. they are 'diploid'). The differential rate of formation of an enzyme or other gene product (i.e. its cell content) is, other factors being equal, usually proportional, within fairly wide limits, to the number of copies of its gene per cell.

It is now possible for the number of chromosomes per cell to increase several fold, so affecting large numbers of genes at the same time. In bacteria certain conditions permit an increase in the numbers of so-called 'plasmids' (minute sub-chromosomes which replicate autonomously and to some extent independently of the main, single chromosome) so that instead of the normal one or two per cell there may be up to sixty or so.[3] This, of course, only affects those genes located on the plasmid, but some of these (for instance, the gene for the penicillin-destroying enzyme, 'penicillinase') may be so located, as well as a number of other genes responsible for drug-resistance.[4]

One of the easiest ways by which these plasmids may get out of control and produce more copies than normally is if they are transferred to other, slightly different, types of bacterium. For instance, if a plasmid from *E. coli* (the colon bacillus) transfers itself to another rather similar species (e.g. the *Proteus* bacillus) where, it seems, the regulatory system is not so well adapted as that of the normal 'home' cell, the plasmid may multiply freely for some time and produce many copies in a single cell.[5]

Another, rather similar, possibility is for a particular bacterial gene to get 'hooked-up' on to a bacteriophage ('phage') so that it replicates as part of the phage chromosome when infecting a particular bacterial cell, producing many copies which form their specific product in large amounts before the cell ultimately dies from its viral 'disease'.[6]

A third possibility for multiple copies of certain genes occurs under certain conditions promoting 'dichotomous' replication[7] of the bacterial chromosome (see Figure 7 in Hayes, p. 50). This gives rise to extra copies of those genes which are near the point where the dividing chromosome starts to replicate.

QUALITATIVE CHANGES

The theoretical range here is enormous. There may be only minor alterations from normal, involving one or a few nucleotides, including repairs to chance defects and restoration to normality. At the other extreme there may be large 'deletions' (complete loss of whole sections of the genome) or the acquisition of completely new genetic material from elsewhere (i.e. from other species of organism).[8] Small alterations to existing genes may nevertheless cause big changes in the properties of the enzymes they produce. It is, indeed, the selected sequence of such changes that must have occurred during the long course of time during which one enzyme gradually evolved into another.

It is also possible for 'mixed' or hybrid enzymes to be produced by the fusion of a part or the whole of a gene to another in appropriate matings between different species and so, in principle, for a new enzyme to arise which could combine the properties of its two 'parents'.[9] Similarly, new combinations of types of enzyme – a new enzyme repertoire, so to speak, might be developed from exchange or accretion of genes during interspecific conjugation.

Mechanisms of genetic modification

We shall now consider the principal phenomena, one or all of which may be involved in the various processes of genetic modification and discuss certain specific instances in greater detail.

Essentially they consist of four basic processes that are of fundamental importance in manipulative genetics.

1. *Random mutation*

Small, usually single, nucleotide changes in long sequences of linked nucleotides (i.e. DNA) involving one base changing to another, or additions or subtractions of bases, are occurring all the time spontaneously (Figure 1). Under the influence of mutagens (chemicals such as mustard gas; physical factors such as ultra-violet light and other types of irradiation) the mutation rate can be very greatly increased (several hundred times). Although not always strictly random, the *types* of change (in relation to the function of the gene product) cannot yet in general be controlled, so that the isolation of particular mutants must depend upon specific selection from a very large number.

```
            A G G C T C A A C G
  . . . .   | | | | | | | | | |  . . . . .            (A)
            T C C G A G T T G C
NEIGHBOURING→ ←—NORMAL GENE—→ ←NEIGHBOURING
   GENE                              GENE


            A G A C T C A A C G
  . . . .   | | | | | | | | | |  . . . . .            (B)
            T C T G A G T T G C
                ↑
            BASE-PAIR SUBSTITUTION


            A G C T C A A C G
  . . . .   | | | | | | | | |  . . . . .             (C)
            T C↑G A G T T G C
            BASE-PAIR DELETION


            A G G A C T C A A C G
  . . . .   | | | | | | | | | | |  . . . . .         (D)
            T C C T G A G T T G C
                  ↑
            BASE-PAIR ADDITION
```

Figure 1 'Minor' mutations: (A) *normal gene;* (B) *substitution of one base-pair for another;* (C) *loss of one base-pair;* (D) *addition of one base-pair.*

Mutations are occurring all the time in all types of cells, including human, but most are eliminated because they kill the cells in which they arise or their effects are masked by a preponderance of normal cells so that we are not conscious of their occurrence.

Specific mutants can be selected very easily with bacteria because such a large number can be handled. It is a simple matter, for instance, to obtain specific, fully derepressed mutants, which have lost the switch-off control of the synthesis of one particular enzyme which may be of special interest. Such mutants are probably put at a disadvantage under natural conditions, but can be very useful as biofactories of the enzyme for human needs.

2. Recombination

Certain enzymes exist in every cell, usually under rigid control, that can cut DNA chains at various points while there are other

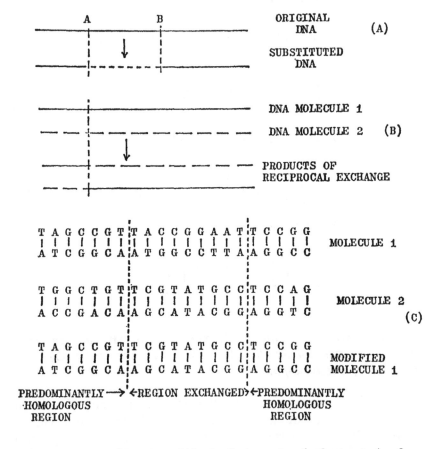

Figure 2 *Recombination:* (A) *single insertion (substitution) of one DNA stretch (sequence of nucleotides) for another;* (B) *reciprocal exchange;* (C) *type of homology between the two pieces of DNA associated with the juxtaposition required for genetic exchange. The mechanism by which homologous sequences in different DNA molecules recognize each other is not yet understood.*

enzymes that can join up the ends again – often so that certain stretches of DNA are replaced by other stretches and positions of genes changed in the chromosome (or moved from one chromosome to another). This is a sort of shuffling process included in the term 'genetic recombination' (Figure 2A, B). Such recombination will normally occur only when the recombining DNA molecules have stretches similar ('homologous') in their nucleotide sequences (Figure 2C).

Such homology is so frequent, at least over small stretches within, or between, genomes of individuals of the same species that reassortment of genetic information can often occur (unless specifically inhibited) both between genes and within genes (especially in bacteria). It can be responsible for major changes in metabolic behaviour: additions or subtractions of enzymes and other vital characters such as pathogenicity and virulence, the ability of groups of genes to replicate autonomously, alterations in rates of formation of gene products and qualitative changes in enzyme properties. The process can be stimulated (non-specifically, as with mutations) by a variety of chemical and physical factors.

Homologous stretches of nucleotide sequences are shorter and rarer between more distantly related organisms, so that recombination between individuals from different species is much less frequent, though it may still occur.[10]

Recombination of genetic material in the form of co-existence and co-ordinate replication of separate chromosomes, from quite different species, within the same cell (or even within the same nucleus) without physical union of DNA appears to be somewhat easier to achieve than physical recombination within a chromosome. In many instances it can result in a similar sort of hybrid cell, though segregation of the elements from the two different sources is rather liable to occur.[11]

3. *Mobilization, transfer and integration*

Genes often pass from one cell to another. This happens most frequently between individuals of the same species by direct contact (e.g. sperm to egg in higher organisms) in the normal process of sexual reproduction.

It is also possible for genes to pass between cells of different species, naturally or by artificial coaxing. This is much less frequent and often only achieved with difficulty and under special circumstances, but is more interesting in relation to genetic manipulation since it is by such means that really new combinations of characters, and even basically new characters themselves, may be constructed.

In general, transfer may be by direct cell contact (through special apparatus coded for by the group of transferred genes itself), by penetration of 'naked' genes (in the form of pure DNA or whole chromosomes) or by facilitated transport, as a sort of stowaway, within the coat of a virus particle having a special affinity for, and apparatus for penetrating, the recipient cell.

The hazards facing 'abnormal' (i.e. inter-specific) gene transfer between cells are numerous. The DNA, if unprotected, is easily destroyed by enzymes or physical inclemencies outside the cell; it has physical barriers to pass if it is to penetrate the cell envelope and the recipient cell may often have developed specific 'protective' devices, operative against foreign nucleic acid from another type of organism, which can prove very effective in destroying the 'invader'.

Finally, even if the intruder gets inside intact, it must still be able to fit in to the strange cell's metabolism, produce its own enzymes and find a way of replicating itself.

There are, however, a range of techniques by which it is often now possible to promote such an intrusion and help the invading nucleic acid to get properly established in its new environment, though it will be possible to refer to only a few of these in this article.

4. *Replication*

Pieces of DNA, otherwise intact and functional, are not automatically endowed with a self-replicating ability, even in an otherwise receptive environment, however long they may be. They must possess, or obtain by recombination, specific genes responsible for initiating replication.

No modification of an organism can be considered *genetic* unless the modification is perpetuated, indefinitely or for a very long time, through successive generations. In other words, new genes introduced into another cell must be capable of faithful replication in order for the change to be heritable.

There are two main ways by which this occurs. The incoming piece of nucleic acid may subsequently be integrated physically into an existing piece of nucleic acid (chromosome or plasmid or virus) that already possesses the ability to be replicated (by virtue, probably, of a small piece of DNA at the end of the chromosome, or at a particular point in its structure if circular, that functions as the sole origin of replication); or it may already possess such an 'initiator of replication' itself, enabling it to replicate forthwith on its own as soon as it is accepted into the recipient cell. In the first instance its pace of replication will be automatically co-ordinated with the rest of the chromosome; but in the latter its replication must somehow be controlled to keep pace with cell division – neither too fast (or it will kill the cell) nor too slow (or it will be lost rapidly from the growing cell population).

Rarely it may be possible for a new genetic element (incapable of autonomous replication) to be introduced into a strange cell where there is normally little chance of automatic integration into an existing self-replicating structure (because of too distant relationships between the species). In such a case integration may occur after previous recombination, under other circumstances, between the new element and a small portion of the future recipient's genome (or another element closely related to the recipient).[12] This may subsequently allow sufficient 'recognition' by the homologous portion of the recipient genome to allow integrative recombination with the latter after its intrusion into the cell. Alternatively, a gene or genes from elsewhere, able to initiate replication (usually a virus or part of a virus, or a transmissible plasmid such as the sex factor F^1 is most frequently involved in this process) may be similarly recombined with the new element, allowing it to replicate spontaneously after introduction into the recipient cell.

The processes of mutation, recombination, transfer and replication described above concern what is in effect an information tape of great length, in some ways analogous to a cinematograph film of which many faithful copies will be made for distribution all over the world. The editing process, which could be compared with the end-result of deliberate attempts at genetic reconstruction, may involve many manipulations with the original copy. These will include the removal of small blemishes and minor alterations as well as chance damage to individual frames (mutations), cuts (deletions), alterations to the sequence of small runs (translocations), and interpolations from a completely different film perhaps shot many years previously (introduction of genes from a different species), involving both cuts and rejoins (recombination). It must be capable of being rolled up and transported in containers to where it is needed (mobilization and transfer of genes from one cell to another) and must have the final seal of the producer's approval (initiation of replication) in order to be passed for copying in the workshops before general release (replication).

In at least one very important respect, however, this analogy is false because all the genetic modifications described above occur only very rarely indeed amongst the population as a whole. Successful extraction of the cell line possessing a specifically modified genetic constitution depends therefore upon the application of an environment (usually a specific growth medium and/or physical factors) by which this variant is selectively favoured (i.e. can multiply at the expense of the normal cells or other variants that may be present). With micro-organisms this is usually not difficult because so many of them can be made available so rapidly. With larger plants or animals it may be a fundamental obstacle to many types of genetic reconstruction, because it is seldom possible to effect a desired genetic change in more than a very small proportion of individuals.

Applied molecular genetics in bacteria

Short-term applications

Molecular genetics has already been applied in microbiology, although so far largely empirically. Examples range from the use, by the pharmaceutical industry, of artificially induced mutants of micro-organisms, forming abnormally high quantities of antibiotics, to the isolation of derepressed bacterial mutants for the production of enzymes such as penicillinase[13] (for treatment of penicillin allergy and blood culture diagnosis of septicaemia).[14]

One particular project now being launched concerns the exploitation of specially constructed strains of bacteria for the economic production of various types of large molecule which cannot yet be chemically synthesized without enormous expenditure of time and money. Possible examples to be considered in such a project would be the production of complex vitamins (such as B_{12}) in cell-free bacterial extracts, nucleases in medicine and surgery for solubilizing sticky exudates,[15] asparaginase[16] as a possible attack on leukaemia, enzymes for the inter-conversion of cortisone-type hormones[17] and a whole range of other enzymes for research and quantitative laboratory assays of complex organic molecules.

The broad aim is to employ newly developed techniques for increasing the number of gene copies in strains where the gene has already been made to work maximally by isolation of fully derepressed mutants (as previously described).

There are several workable possibilities for doing this, the most promising being the following.

Plasmid link-up. Variants showing recombination of the gene (or genes) determining the structure and synthesis of a particular enzyme with genes of an autonomous extra-chromosomal plasmid can be artificially selected. The relevant gene in these variants is found to be physically part of the plasmid.[18] This recombinant gene-carrying plasmid can then be transferred to another type of bacterial cell. As previously pointed out, it often happens[19] that plasmids so transferred to a 'strange' host are not so well controlled as in their normal host environment and multiply more freely, producing many copies per cell. The enzyme-coding gene which has become part of the plasmid is replicated along with the the other plasmid genes and will have the same increased number of copies. The rate of enzyme production per cell is increased proportionally.

In this sort of problem, as in some other types of genetic manipulation with bacteria, it is important for a particular gene to be able to move from its normal position on the chromosome not only on to a plasmid but to other locations on the chromosome itself ('translocation'). The procedure involves the selective isolation of very rare cell lines arising from the integration of a plasmid – having 'episomal' properties[20] and recombined with the gene in question – at identifiable chromosomal locations different from the normal site. Integration at abnormal sites will occur (very rarely) under conditions where the usual site has been intentionally eliminated by a deletion.

Essentially this trick depends upon using, as a gene-carrier, a genetic element that can move around from one position to another on the chromosome. Limited success has already been achieved by this technique in *E. coli* with genes for the enzymes responsible for splitting lactose[21] and there seems to be no reason why it should not be extended to other genes.

Phage link-up. Association (again by translocation procedures plus appropriate selection similar to those already described above) of the desired gene with the genes of a bacteriophage can occur at a stage when the phage is in what is referred to as a pro-virus ('pro-phage') state[22] – that is, behaving as a group of genes on the bacterial chromosome or multiplying independently, but co-ordinately, in the cytoplasm.

Several types of pro-phage can be induced (e.g. by ultra-violet light and other factors) to escape from the chromosome and/or its controlling influence and resume rapid multiplication autonomously in the cell cytoplasm, ultimately bursting out of the cell they have so killed as fully fledged virus particles. If the enzyme-coding gene has been appropriately combined with the phage genes (without displacing too many of them) it will be replicated along with the virus. In order, however, to allow the enzyme itself to be formed properly in proportion to the increased number of gene copies available, it is necessary to isolate a phage mutant of a type[23] that is defective in its virulence and does not kill the cell too rapidly. Several fold increase in yield of certain enzymes has already been achieved in *E. coli* by this particular manoeuvre.[24] There seems to be no reason why these procedures could not be extended to other enzymes, the main limitations probably being those connected with translocations, selective isolation procedures for which may be difficult to institute for certain genes reluctant to recombine with specific pro-phage genes at a particular position on the chromosome.

Dichotomous replication of chromosomes. Multiple replication points on the bacterial chromosome (see above and Figure 7 in Hayes, above) can be induced by transferring cells to a specially rich medium and allowing them to grow at maximal rate.[25] This may induce three or more replication forks, instead of the usual single one. By the use of certain mutants and particular growth conditions, it is possible to stimulate cells to over-produce their DNA (in relation to the rest of the cell matter) which they may do by increasing the number of replication points up to as many as fifteen.[26]

It may further be necessary to translocate the desired gene to a point near the origin of replication in order that it may benefit maximally from the increased number of copies. And that can be done by the same sort of recombination plus selection techniques described above.

Artificially synthesized genes. Finlly, it is now beginning to be possible to synthesize whole genes in the test tube and so construct a unique polynucleotide sequence exactly as required. A very small artificial gene for a rather special type of so-called 'transfer-RNA' molecule (not, in fact an enzyme, but somewhat analogous) has already been so produced,[27] although nobody yet knows whether it can be made to work. It has, however, been claimed possible to tack on to a virus genome a very simple repetitive nucleotide sequence (AAAAAA . . .), coding for a sequence of the amino acid lysine, and show that cells infected with this modified virus will produce a sequence of lysine molecules strung together ('polylysine') and that the modified virus breeds true, retaining its modification and continuing to stimulate the production of polylysine.[28] There is some reason, therefore, for believing the procedure would work in principle in all living cells.

Again, in principle, *any* sequence of nucleotides (and therefore theoretically any gene) could probably be constructed chemically – given enough time. But nobody yet knows how to make them work properly or even what sequence would be needed to produce a protein with a given function. This is simply because the chemistry of existing proteins is so little understood (with very few exceptions) that no generalizations are yet possible to allow scientists to design genes which would form enzymes with a completely new (i.e. hitherto unknown) function.

It might, for instance, be considered desirable under certain circumstances to be able to destroy the artificial, and inconveniently stable, pesticide DDT by biological means. And there must be many holidaymakers who wish there were soil microorganisms capable of decomposing plastic containers as easily as they can 'ordinary' refuse.

All that can be envisaged for the immediate future would be the chemical synthesis of genes for known enzymes – and that is still some way off itself – let alone the practical possibility of getting them to form their enzymes to any advantage over living organisms. However, very small artificial modifications – additions or subtractions – at the ends of gene groups (such as small viruses), as the example above indicates, are more plausible possibilities that are already being explored for sealing DNA circles and joining genes together in more grandiose recombinational processes that will be discussed in greater detail later on.

Longer-term possibilities

Those of the above mentioned possibilities for genetic engineering in bacteria that have already been tested mainly involve only quantitative alterations in the differential rate of formation of certain enzymes. It is important, however, to realize that some of these processes may lead to an increased production of several hundred or even thousand fold over that available in the wild-type organism – and all this in the same medium without any increased cost of production. In effect, it is a method for getting bacteria to increase the proportion of a particular material they produce from the 'normal' (say 0.1 per cent) of total mass up to 20 per cent or so.

Longer term, so far little tried and altogether more speculative (but not implausible) possibilities that are now worth considering involve construction of 'new' types of cell with new genetic constitutions by introduction of foreign DNA from other organisms, in order to combine two or more specially useful characters from different kinds of cell. For instance, it might be possible to combine the rapid growth of *E. coli* in very cheap medium with the production of particularly valuable enzymes (such as those synthesizing vitamin B_{12}) only formed naturally by other, slow-growing or exacting organisms.

Another analogous possibility would be to introduce the genes for enzymes capable of using atmospheric nitrogen gas as a source of nitrogen for growth (as are possessed by the N-fixing bacteria attached to the roots of clover and other plants), enabling them to grow without nitrogen compounds in the soil and/or carbon dioxide by photosynthesis, into organisms useful for specific purposes (as already described) which normally need more complicated and expensive sources of carbon and nitrogen.

The theoretical possibilities here seem almost limitless. Perhaps one of the most valuable of all would be the continued use of cheaply and rapidly growing bacteria acting as artificial hosts for the replication of animal viruses such as vaccinia or poliomyelitis (for use as vaccines in human immunization) which now have to be grown expensively in animals or animal tissues. This is not, perhaps, so far fetched as it may seem at first sight. Indeed, a few years ago there were claims that this sort of process had been achieved.[29] The fact that it has not been possible to repeat the experiments in other laboratories does not necessarily mean either that the original experiments were false or that ways may not soon be found of putting such possibilities into controlled and repeatable practice – because they are possible examples of the *sort* of phenomenon which sooner or later is almost certain to be proved feasible.

Short-term applications and long-term possibilities 75

Another interesting possibility would be the use, in those programmes for biological control[30] now being promoted by Unesco, of suitably modified strains of bacteria rather than the normal wild type, in order to offset dangers of too great an ecological imbalance and/or to improve effectiveness during short-term applications.

The extent of 'mixed marriages' possible between different species (especially in the form of different types of nuclei multiplying in the same cell, or even sets of chromosomes – from birds and mammals – within the same nuclei[31]) seems to be continually expanding.

So far as bacteria are concerned, what especially is needed would be knowledge of the types of compatibility physiologically possible and techniques for introducing 'foreign' genes through the external cell barriers, as well as procedures for promoting and selecting the probably extremely rare recombination events that may thereby be facilitated.

Measures for fusing different types of bacterial cells, perhaps by weakening their cell envelopes (in a manner analogous to that used in fusions of cells from different families of vertebrates) should lay the foundation for co-operative interactions and even perhaps more permanent recombinations between two species of genome.

Applications to man

It is inevitable that techniques of genetic reconstruction found possible, or thought plausible, for bacteria should be considered for possible extension to human beings. Indeed it has very recently been claimed possible to transform (genetically) isolated skin cells from albino mice to pigment (melanin) production by treatment with DNA extracted from all types of cell (liver and other organs) from a normally pigmented mouse.[32] It also seems possible to transfer genes from one mouse cell to another by the use of an animal virus (polyoma) which acts as a carrier in what appears to be a manner similar to that operating with transducing phages for bacteria although actual recombination has not yet been demonstrated.[33]

In principle, then, and as expected, what can happen with bacteria will probably often be possible with the cells of higher organisms. The way is, in a sense, open for genetic intervention in man. In practice, however, the situation is very different.

Extravagant fantasies of genetically re-fashioned babies growing in test-tubes or the directed construction of human genotypes with giant intellects and no feelings, or all brawn and no mind, take no account of reasonable scientific possibilities, let alone social acceptability. Cloning of individuals, by substituting the normal nucleus of a fertilized egg with that from an adult cell and subsequent replacement in the womb, would produce an exact genetic copy of the adult donor and theoretically this could be repeated many times with the same donor. It is *scientifically* more plausible because it has been achieved (although under much simpler conditions) with frogs[34] and may soon be possible with mammals. Facilitated inter-specific fertilization of egg with sperm (for instance between man and ape) is not at all inconceivable and in principle only one step removed from mating a horse and donkey. But both procedures have monstrous implications.

Of course all things (or at least many very unlikely things) are conceivable. But if we are to worry about real dangers in the immediate future, we would do better to consider the possibility of blowing all ourselves up by the huge number of H-bombs we possess, which is many times more likely than producing the 'little green man in a test tube', and try to do something about that.

Even from a purely biological point of view, genetic engineering in man is an altogether different problem from that in bacteria. Genetic deficiencies (i.e. those arising from defects in the genome itself) are extremely common but most are unnoticed because the fertilized egg (zygote) dies from the condition long before birth. Those that survive may face a distressing life – they may be imbeciles, cripples, or suffer from haemophilia, cystic fibrosis, etc. – and we should greatly like to be able not only to cure them but to ensure that their children and their children's children are normal. It has been estimated that possibly up to 15 per cent[35] of births show *some* evidence of inherited disorder though this may often be relatively trivial.

Many, many more are themselves normal, but carry the defective gene in an inactive state (or with only minimal untoward effects) but if married to carriers of the same defective gene, run the risk of producing seriously handicapped children.

There are many types of genetic deficiency now known in man (most of them fortunately rare) where it has been possible to identify the defect in the sense of knowing what specific enzyme has 'gone wrong', though little may be known about the gene itself.

A defective enzyme may mean that a particular vital biochemical reaction does not take place and an essential body constituent is not manufactured in the body. Theoretically, then, the deficiency might be made good by supplying the missing ingredient continuously during the patient's life, in the food or by injection.

A more long-lasting cure would (theoretically again) be to supply the normal enzyme itself, though (except in the case of enzymes operating exclusively in the bloodstream) problems of getting it into the cells intact might prove to be almost as difficult as introducing a gene. It is possible, for instance, to treat the disease diabetes (*not* usually due to a genetic defect) with an essential protein (insulin) with great success, but of course it has to be given continuously throughout life. With a true genetic defect, a radical cure might be achieved were it possible to introduce the relevant piece of normal DNA and promote its proper integration into a sufficiently high proportion of the cells that should normally be forming the enzyme.

It is the last condition which may be the main problem. In bacteria repair is easy to demonstrate, but it depends upon artificial selection (a sort of amplification) of the relatively small proportion of cells so modified. In man and other higher organisms however, it may be very difficult to get such repair effected, if it can be managed at all, in more than a minute percentage of the cells that need to be modified.

Nevertheless, in the case of an enzyme deficiency, it may not be necessary to repair more than (say) 5 per cent of the cells, since 5 per cent of normal enzyme activity could well be enough to supply quantities of the missing factor sufficient to prevent symptoms. Moreover, there are possibilities involving the use of relatively harmless viruses as gene carriers which might allow transfer of the normal gene to a high proportion of the relevant deficient cells.

One specific case which has attracted attention recently concerns the rare genetic disease argininaemia in humans, where an essential enzyme needed for removing excess arginine from the bloodstream (arginase) is missing, or defective, because of a genetic 'error' in the gene responsible for its formation.[36] Arginine is a normal amino acid required as a building block in protein biosynthesis, but too much of it is poisonous, causing mental retardation, epilepsy and often death. In this instance, therefore, the enzyme defect results in excess production of a potentially toxic substance rather than a simple deficiency of an essential nutrient factor, as discussed above.

Now it so happens that a virus normally infecting rabbits (the 'Shope' papilloma) can also infect man but without apparently causing any harm. This virus contains a gene for arginase so that conceivably there could be a chance of treating or at least modifying the disease in man by infection with this virus. In the rather unlikely event of any sort of success being achieved in this case, one of the next steps might be to try and 'stitch on' specific genes to harmless man-infecting viruses by artificial recombination, using isolated enzymes or even purely chemical bonding processes for the purpose. It might then be possible to make use of the invading and self-replicative properties of the virus for carrying normal genes into defective cells and so repairing a wide range of genetic diseases.[37]

All this is still a very long way off, but it is at this sort of level that it may now be reasonable to envisage practical future possibilities in human genetic intervention.

Yet, in a sense, these interventions are not really properly genetic at all. At best, the repairs effected may be passed on during multiplication of the liver or kidney cells etc., that require repairing in the individual himself. But his children will retain the defect.

In order to effect a truly heritable cure, it will be necessary to repair the defect in the genome of the germ cells themselves: the sperms and eggs and their precursors. And it will be necessary to do this in *all* (or at least the overwhelming majority) of the germ cells, not just the 5 per cent which might be sufficient for an individual ('phenotypic') cure.

At the moment there is no foreseeable way by which this could be achieved, although, again in principle, it need not be considered intrinsically impossible. More serious, however, than this difficulty in seeing how a true genetic cure might be obtained, is the whole problem of *any* attempt to modify the 'germ line'. Far too little is known about possible dangers in this field to permit manipulations that might cause permanent distortion of generations of human beings or condemn them to sterility.

It may be considered permissible to take *some* risk in attempting to treat dangerously ill patients in the hope of a phenotypic cure, but manipulations intended to affect the next generation are in a different class. Even interference with genes of non-germ line ('somatic') cells (e.g. of liver) by virus infection, intended to cause repair of their genome, may run the risk of a 'spill over' into the germ cells themselves and should therefore be approached with great caution.

It could be argued that the best hope eventually of permanent genetic cure of hereditary defects would have to be applied locally at a very early stage of gestation, when only one or a very few cells are present in the embryo: that is, at a time when all the cells including the future germ-line *might* be repaired. Then, in the event of failure or serious damage (*if* it could be detected) the worst outcome would be a miscarriage and not a crippled human being.

But most people would rightly conclude that these sorts of considerations only go to show how premature it still is to consider such manipulations as practical possibilities in the present state of our knowledge.

What might be the position twenty years from now, however, no one can say.

Bibliography

References

1 Hayes, W., Figure 8 (p. 52 above).
2 Rogers, S. (1970). *New Scientist 45* (29 January) 194.
3 Rownd, R. (1969). *J. mol. Biol. 44*, 387; Kontomichalou, P., Mitani, M. and Clowes, R. C. (1970). *J. Bact. 104*, 34.
4 Novick, R. P. (1969). *Bact. Rev. 33*, 210; Watanabe, T. (1967). *Sci. American 217*, no. 6, 19.
5 Rownd, *op. cit.*; Kontomichalou *et al., op. cit.*
6 Smith, J. D., Abelson, J. N., Clark, B. F. C., Goodman, H. M. and Brenner, S. (1966). *Cold Spr. Harb. Symp. Quant. Biol. 31*, 479.
7 Oishi, M., Yoshikawa, H. and Sueoka, N. (1964). *Nature 204*, 1069; Yoshikawa, H. and Haas, M. (1968). *Cold. Spr. Harb. Symp. Quant. Biol. 33*, 843.
8 Pollock, M. R. (1969). *Progr. Biophys. molec. Biol. 6*, 273.
9 Sarkar, S. (1966), *J. Bact. 91*, 1477; Yourno, J., Kohno, T. and Roth, J. R. (1970). *Nature 228*, 820.
10 Pollock, *op. cit.*
11 Ephrussi, B. and Weiss, M. C. (1969). *Sci. American 220*, no. 4,

26; Engel, E., McGee, B. J. and Harris, H. (1969). *Nature 223*, 152.
12 Watanabe, T. and Ogata, C. (1965). *J. Bact. 91*, 43; Demerec, M. and Ohta, N. (1964). *Proc. natl. Acad. Sci. Wash. 52*, 317.
13 Pollock, M. R. (1957). *J. Pharmacy & Pharmacol. 9*, 609.
14 Becker, R. M. (1960). *Practitioner 184*, 447.
15 Johnson, A. J., Ayvazian, J. H. and Tillett, W. S. (1959). *New Eng. J. Med. 260*, 893.
16 Adamson, R. H. and Fabro, S. (1968). *Cancer Chemotherapy Rep. 52*, 617; Ho, P. P. K., Frank, B. H. and Burck, P. J. (1969). *Science 165*, 510.
17 Charney, W. (1969). *New Scientist 43* (25 September) feature section, p. 10.
18 Hayes, W., above, p. 56 and Figure 10.
19 Rownd, *op. cit.*; Kontomichalou *et al., op. cit.* (see n. 3).
20 Hayes, W., p. 56 and Figure 10.
21 Beckwith, J. R., Signer, E. R. and Epstein, W. (1966). *Cold Spr. Harb. Symp. Quant. Biol. 31*, 393.
22 Gottesman, S. and Beckwith, J. R. (1969). *J. Mol. Biol. 44*, 117.
23 Signer, E. R. (1969). *Nature 223*, 158.
24 Brammar, W. J. and Burdon, M. G. (1970). Private communication.
25 Oishi *et al., op. cit.* (see n. 7).
26 Yoshikawa and Haas, *op. cit.* (see n. 7).
27 'Monitor' (1970). *New Scientist 46* (11 June), 510.
28 Rogers, *op. cit.* (see n. 2).
29 Abel, P. and Trautner, T. A. (1964). *Vererbungslehre 95*, 66; Bayreuther, K. E. and Romig, W. R. (1964). *Science 146*, 778; Ben-Gurion, R. and Ginzburg-Tietz, Y. (1965). *Biochem. biophys. Res. Comm. 18*, 226.
30 Royal Society (1970). 'Biological Control'. The U.K. contribution to the proposed international organization for Biological Control. Brit. Nat. Cttee for Biology (Biological Control Sub-Cttee); Wilson, F. (1970). *New Scientist 48* (8 October), 72.
31 Ephrussi and Weiss, *op. cit.;* Engel *et al., op. cit.* (see n. 11); Harris, H. and Cook, P. R. (1969). *J. Cell Science 5*, 121.
32 Glick, J. L. and Saline, A. P. (1967). *J. Cell Biol. 33*, 209; Ottolenghi-Nightingale, E. (1969). *Proc. natl. Acad. Sci. Wash. 64*, 184.
33 Osterman, J. V., Waddell, A. and Aposhian, H. V. (1970). *P.N.A.S. 67*, 37.
34 Gurdon, J. B. (1968). *Sci. American 219*, no. 6, 24.
35 Lederberg, J., quoted in *New Scientist* (1970) *46* (18 January), 564.
36 Rogers, *op. cit.* (see n. 2).
37 *Ibid.*

General reading

Clowes, R. (1967). *The Structure of Life*. London, Penguin.
Leach, G. (1970). *The Biocrats*. London, Jonathan Cape.
Nass, G. *The Molecules of Life*, translated from the German by David Jones (1970). London, Weidenfeld & Nicolson.
Paterson, D. (ed.) (1969). *Genetic Engineering*. London, B.B.C.

Discussion

Yon When you embark on a series of experiments on the modification of bacterial genes, how important do you think it is to consider, before actually doing the experiments, the possible effects of introducing into the environment new organisms which may have tremendous selective advantages over their competitors, e.g. an *E. coli* which can fix atmospheric nitrogen.

Pollock I rather suspect that people have been misled by science fiction into believing that it is likely. This is possible. One would treat 'synthesized' organisms in the same way as one would treat pathogenic organisms we are already accustomed to using in the laboratory. Practically all mutants and recombinants are reconstructed organisms in the sense that they are produced by the incorporation of, or slight modification to, already existing genes. And it is frequently only a question of selecting ones that are already there. I believe most geneticists wouldn't therefore think that the danger you imply was at all great.

Salaman If we are opposed to advances in microbial genetics being used in the development of virulent organisms for use in biological warfare, what should we do either as scientists or as ordinary citizens to influence government policy on these developments? Should we not attempt to influence the individual scientists working in government establishments, e.g. by excluding them from scientific societies and by making it impossible for their non-classified work to be published in scientific journals?

Beckwith I think that the trouble with boycotts and sanctions against people doing biological warfare research is that it suggests the people who are imposing the boycott are somehow holier and not involved at all, whereas in fact we are all involved. Having said this, obviously there are also measures which we should take to discourage people from doing biological warfare research.

S. Rose I think the reasons biological warfare is perhaps less important than many people have feared is precisely that in fact biological warfare is probably not a very useful, efficient or sensible weapon for either major or minor powers to develop; in contrast chemical weapons can have a much more substantive tactical and strategic use, and they are of course used in Vietnam in a way which biological weapons have not yet been.

Having said that, I do not think that one can simply say, when discussing boycotts, as Jon Beckwith did, that we are all in this together. There is a large gap between the work that he does in his basic laboratory and the work that is done at institutes like Fort Detrick in the United States, or Porton in this country. This is a gap involving development, and I think it is perfectly proper to call for the cessation of secret research at institutions like Detrick and Porton. I should be happy to see a situation where, until Porton's work is non-classified and totally public, papers by Porton scientists are not accepted at international or national meetings, and learned and scientific societies state that they do not regard these individuals as scientists within the accepted scientific community but, in a phrase that was coined elsewhere, as 'soldier-technicians' whom we may regard with proper disparagement. I think there is a strong case for such a holding measure until institutions like Porton cease to do secret research.

Simon I should like to disagree with Professor Rose because, although institutions like Porton must act in a socially responsible way, I think to suggest that they should publish their results would have a deleterious effect on the world community. For example they would then be publishing details of aerosol distribution mechanisms, the sort of thing which small countries who wanted to use this cheap and semi-effective weapon would want to know.

S. Rose If, as Porton claims, its work is for defensive purposes, genuinely civil defensive purposes, there is no need to do the sort of research on aerosol distribution that you are talking about.

Monod I think I agree largely with what Rose said; I feel personally that even though biological warfare may not be a very effective way of waging war, it is still bad for the soul of the scientific community that there are people working in this area who are still recognized as members of the community.

Part three Human genetics and reproduction

Six Social effects of research in human genetics

Geoffrey Beale
Royal Society Research Professor,
Institute of Animal Genetics
University of Edinburgh

In order to prepare this article I tried to classify various types of genetical research into two groups which may be called 'beneficial' and 'sinister'. Such a classification, though of course extremely arbitrary, and meaningless for the mass of 'neutral' work of a purely scientific or academic nature, does serve to highlight possible good and bad consequences of our activities as research scientists. Looking back over the work of the past fifty years or so, however, I find it rather difficult to name any particular genetical discoveries which have produced positive harm; on the contrary many have been of considerable benefit. Since I intend later on to say something about possible dangers arising from work now in progress or planned, it is worth while to list some of the past benefits, so that we can form some sort of balanced view.

First, therefore, it may be useful to say a few words about each of the following topics: (1) human biochemical genetics; (2) human chromosomal variants; (3) the effects of radiation and mutagenic chemicals.

Human biochemical genetics goes back to the work of A. E. Garrod, early in this century, on the condition known as alcaptonuria. This, being very rare, and producing relatively minor clinical effects, is of no great social importance, but served to introduce the totally new concept of 'inborn errors' of metabolism. We now know about a considerable number of other inborn errors,[1] such as phenylketonuria (PKU) and galactosaemia, which produce severe effects – insanity, gross physical defects or early death, in an appreciable number of people. Thanks to the combined efforts of biochemists and geneticists it is possible to alleviate, to some degree, the disastrous consequences of the activities of the genes controlling some of these 'errors'. This is

Social effects of research in human genetics 85

done by modifying the diets of babies, omitting or reducing some food component such as the amino acid phenylalanine in the case of PKU or the sugar galactose in the case of galactosaemia. By such treatment some people can live a much more tolerable life than would otherwise be possible, though admittedly there is no chance of a 'cure' of the basic genetic abnormality.

As regards more 'permanent' repair, i.e. the elimination of a genetic defect altogether from individuals or populations, present genetic knowledge and technique is of little or no help, except to advise persons containing deleterious genes not to have children. I will not dwell in these eugenic problems, which will be discussed by Penrose.

Turning to chromosomal variants in human beings, it is of interest to point out that the study of chromosomal abnormalities was at one time considered to be a rather academic exercise, and no one could have foreseen the implications for mentally defective people of the studies made forty years ago on trisomics (cells having one extra chromosome) in plants. Now it is clear that several important human aberrations are caused by such chromosomal irregularities. One of the commonest is Down's syndrome (Mongolism), due to the presence of an extra chromosome – no. 21 in the human set of 23 pairs – either free or attached to another chromosome.

There are also a number of abnormalities in human beings involving the sex (X and Y) chromosomes, some having relatively minor effects (XXX), others more pronounced (XO, XYY). I will say more about the type called XYY later. All these chromosomal variants can be detected by microscopic examination of a baby's cells, or by amniocentesis as described by Fairweather (p. 102).

Unfortunately we know practically nothing about the ultimate causes of the irregularities in chromosome behaviour responsible for these conditions, though as discovered many years ago by Penrose, the age of the mother is an important factor in some cases. Naturally it is of considerable value to be able to recognize these errors at a stage at which abortion is a possibility, so that birth of a severely defective person can be prevented.

The above-described gene and chromosome changes occur spontaneously in human populations, in the absence of man-produced environmental stimuli. However, certain artificial agents are liable to affect the genetic material. H. J. Muller in 1927 proved that treatment of the fly *Drosophila* with X-rays resulted in a large increase in the number of mutations occurring, and for many years thereafter continued to issue warnings about the harmful genetic effects of radiation used in medical treatment. These warnings did not have much effect until about 1956, when both the American and British governments set up commissions to enquire into the biological effects of radiation arising from atomic bomb tests. This was a consequence of the widespread public anxiety about these matters at that time.

No direct measure of the effect of radiation on human genes was available then and the position is much the same today, except that work is now beginning with cultured human cells as experimental material. From studies on animals such as mice and *Drosophila* however it is certain that all exposure to radiation is genetically harmful, and should be kept to a minimum. Much greater care is taken about this now, than formerly.

We are also now aware that many chemical substances are liable to produce harmful genetic changes. Most of the substances studied are unlikely to be a serious threat, since they are not usually present in high enough concentrations to affect human reproductive cells. But there are a vast number of substances which have not been tested for mutagenic activity, and which might well be dangerous. An environmental mutagen society has been set up to study this, and it is important that many pesticides, insecticides, food additives and medicinal preparations now in common use should be tested. We shall have to balance up the beneficial and harmful effects of some of these products. As Muller many times pointed out, every additional mutation in human beings produced by radiation or chemical agents adds to the genetic 'load', and sooner or later, possibly after many thousands of years, some unfortunate individual is liable to suffer by physical impairment, mental defect or early death. It is very difficult for us to appreciate this injury which may be inflicted on some individual thousands of years hence, due to some activity taking place now.

What I have said is intended to show that one of the main beneficial consequences of genetical research has been to warn of the possibility of occurrence of harmful changes in the heredity of human beings and to suggest means of avoiding them, or of alleviating them when they do appear. However, one should not forget that positive good has resulted too. The most spectacular example of this is the vast increase in food production which the application of genetics to plant breeding has achieved. Yields of the main world food crops – rice, wheat and maize – have been immensely raised by the application of a simple knowledge of Mendel's laws[2] (as further discussed by Galston). Somewhat less spectacular but still very important gains in animal production, especially with poultry, pigs and milk also owe much to genetic knowledge.[3] No doubt these are primarily economic matters, but indirectly the social consequences must have been very great.

One further example of the application of genetics to human affairs may be mentioned, namely its value in cases of disputed parentage. Genes controlling the ABO, Rh and other blood groups have been used for many years for this purpose, and now so many different blood group systems are known, as well as many genes controlling protein variations, that with sufficient effort it should be possible to establish the exact biological parentage of all of us, if that should be considered desirable.

It is also sometimes said that our knowledge of genetics enables us to regard in a more rational light prejudices about races, social classes and nationalities, but one is not aware of much progress in this matter so far.

Having mentioned briefly some of the past applications of genetics, I now want to ask: What further developments are likely and what will the consequences be? No doubt further progress will be made in ameliorating inborn errors along the same lines as before, and one hopes that defects whose biochemical basis is at present obscure will become well enough understood for remedial measures to be possible. This applies to cystic fibrosis (the commonest serious human genetic defect), muscular dystrophy, haemophilia, Huntington's chorea and many others, and for that matter also to sickle-cell anaemia, which is biochemically clear enough but still a great source of human misery. In a strictly scientific sort of way the problem is straightforward, since by now we must be familiar with all the commoner gene and chromosomal defects in man. Large surveys of the world population have been made and the same gene mutations, the same chromosome variations, occur again and again. It is very unlikely that totally new genetic defects will appear in the future, except much more rarely than those already known.

Biochemical treatment of genetic defects in the lifetime of the affected individuals constitutes a slightly increased burden on the medical services, but one which can easily be borne. It is sometimes thought that such a burden would significantly increase if people with genetic defects were to reach maturity and have children, but it can be calculated that the rate of increase of rare recessive genes in the population brought about in this way would be scarcely perceptible in a few generations.

Another aspect of biochemical genetics which is likely to advance is that concerned with the recognition of persons who are heterozygous for recessive genes, but do not show any of the effects. A number of sex-linked genes, such as those for haemophilia, Duchenne muscular dystrophy, Hunter's disease and others can already be recognized in heterozygous women. This is very valuable, since half the sons of such women, if they were to have children, would be expected to be severely defective.

As for the more radical techniques of direct 'repair' of the genetic material in the germ cells of persons containing deleterious genes – whether by substitution of whole chromosomes or by insertion of small segments of DNA or individual genes by some kind of virus-mediated transduction – I will leave discussion of these exceedingly important matters to others. Likewise the very delicate subject of manipulation of human eggs, sperm and embryos, and the control of sex of children, will be left to Edwards. Even without these topics, however, which pose so many vitally important social questions, I am still left with certain other types of genetical research, which show how harmful social consequences may arise, unforeseen by those who initiate the work.

To illustrate this, we can consider the consequences of conducting large scale surveys of gene and chromosomal abnormalities amongst human populations. The knowledge gained will undoubtedly be of great value to the community, and to some individuals, but other individuals may suffer. We may take as an example the XYY chromosome story described in 1965 by Jacobs and others.[4] As everyone knows, cells of human males have a Y chromosome in place of one of the two Xs of females. Occasionally, males with two Ys occur, and the first examples of this condition were discovered in a maximum security penal institution in Scotland. Some of these men were described as 'dangerous, aggressive, mentally subnormal psychopaths'. Naturally, widespread interest was aroused by the publication of this information, especially when similar findings were reported from other countries. The impression spread around that there was an invariable association between the XYY condition and criminality.

More recently, however, members of the same research group as that concerned with the original finding have obtained clear evidence that XYY men occur quite commonly in the general population, and some of these – probably the majority – have no behavioural disorders whatever. The frequency turns out to be unexpectedly high, about one in seven hundred males, as determined by studies on babies born in two Edinburgh hospitals,[5] or amongst adult males in France.[6] Moreover a large survey of men in various penal institutions in Scotland failed to show any higher incidence there than amongst the general population.[7]

It is very unfortunate that some of the first XYY cases discovered were criminals. The result may be to make other XYY men, if their chromosome type were known, become objects of quite unjustifiable suspicion. People might be unwilling to employ them, and the interest of the police in them might be aroused. Fortunately, the estimated 2,500 XYY males at large in Scotland, who are probably indistinguishable from their 'normal' XY fellow-citizens, have not been identified.

In a general way the same considerations apply to the publication or registration of any kind of adverse genetic information about an individual. It has been pointed out to me that identification of individuals as heterozygotes for known deleterious genes might result in some deprival of liberty to those individuals, who might thereby be dissuaded from marrying the persons of their choice. Personally I find the position of such heterozygous persons somewhat less disastrous than that of the XYY men just described. Heterozygotes involving the commonest harmful recessive gene in human beings – cystic fibrosis – constitute only about one-twentieth of the population. If you were one of them, the chances of your being prevented from marrying a particular person would not be great, and for most other defects the frequencies would be much lower. But others may take a more serious view about this.

A more threatening problem may arise from the proposals now under discussion to record on computers the genotypes of large numbers of people, possibly of whole countries, with the commendable aim of trying to correlate particular genotypes with susceptibility to cancer or other illnesses. In putting such schemes into operation one should bear in mind possible loss of freedom to persons thought to be in one way or another 'defective'. I have in mind restrictions on employment or on travel from one country to another, as well as on choice of a mate. Of course these objections would be less valid if the information were to be kept secret, or if the identity of the persons contributing the information were to be automatically concealed. However, a government is always in a position to insist on having such information. And with increasing education in genetics, the individuals involved might also want to be informed.

This brings me to the final question of whether it is desirable that certain types of research should be prohibited altogether, because of the possibility of 'sinister' exploitation of the results by other people. A particularly sensitive area of research is that relating to intelligence, its possible genetic basis and variation amongst different social classes and human races. Some people take the view that such knowledge is potentially so explosive, so liable to be used to exacerbate racial tension, that it should not be made public at all. Monod has already mentioned this problem but I feel that it is so important and relevant to the theme of this meeting that further discussion is required.

In a recent issue of *Scientific American*,[8] Bodmer and Cavalli-Sforza have an article entitled 'Intelligence and Race', describing the attitude of geneticists to this problem. At the end of their article they come to the conclusion that further research of this kind should be discouraged by cutting off public funds in its support. On reflection I find I am not in favour of this proposal, even if it were possible to make such a ban effective, which is very unlikely. If one considers the particular example of the genetics of IQ, one reason for the emotional argument that goes on is the poor quality of the scientific information available. Looking into the literature on the subject one is struck by three things. First, the factual data are meagre, mainly based on a few twin studies made about thirty years ago. Second, the actual 'measurements' or assessments of IQ are not objective, due to various uncontrollable variables, such as the cultural backgrounds both of those who prepare and conduct the tests, and of those who are the objects of the investigation. Third, genetic analysis of a continuously varying character such as IQ in man is complex and difficult. In view of these three sources of confusion, it is not surprising that there is a lack of agreement about the significance of the findings.

It may be that the problem of intelligence and race is so difficult that it will be impossible ever to obtain the information necessary for a precise genetic analysis. But it seems to me that it would be defeatist to accept this as the final position. New techniques, better criteria for assessing the human intellect, new basic information on the genetics of behaviour, and new analytical methods, will surely become available in the future, and enable us to make more effective approaches to the problem. We shall progress beyond merely asking such questions as whether the IQ of one person is higher than that of another, and what proportion of the difference between them is due to genetic causes. More subtle and penetrating questions will need to be asked. To ban this kind of research now would be tantamount to acknowledging the existence of a fear, based on ignorance, that the racists are right, that some human races are genetically 'inferior' to others. In my view there is no justification whatever for accepting this, and it would be a great error to let the matter go by default. After all, the whole point of scientific research is that one doesn't know in advance what the conclusions will be.

Therefore I do not subscribe to the view that research on genetics and intelligence, or even on racial differences, should be banned. Our present difficulties will not be solved by doing nothing, but by formulating new concepts, asking new questions, and eventually planning new investigations. But I certainly agree that the greatest possible care should be exercised when the time comes to publish the results of such studies.

References

1 Harris, H. (1970). *The Principles of Human Biochemical Genetics.* Amsterdam–London, North Holland.
2 Riley, R. (1969). 'Plant genetics in the service of man', in J. Jinks (ed.), *Fifty Years of Genetics*, Edinburgh, Oliver & Boyd, 27–36.
3 Robertson, A. (1969). 'The application of genetics to animal breeding', in *ibid.*, 37–56.
4 Jacobs, P. A., Brunton, M., Melville, M. M., Brittain, R. P. and McClemont, M. M. (1965). 'Aggressive behaviour, mental subnormality and the XYY male', *Nature 208*, 1351.
5 Ratcliffe, S. G., Melville, M. M., Stewart, A. L. and Jacobs, P. A. (1970). 'Chromosome studies on 3,500 newborn male infants', *Lancet* i, 17 January, 121–2.
6 Noel, B., Quack, B., Durand, Y. and Rethore, M. O. (1969). 'Les hommes 47, XYY', *Ann. Génét. 14*, 223.

7 Jacobs, P. A., Price, W. H., Richmond, S. and Ratcliffe, R. A. W. (1971). 'Chromosome surveys in penal institutions in Scotland', *J. human Genet.* (in press).
8 Bodmer, W. F. and Cavalli-Sforza, L. L. (1970). 'Intelligence and race', *Sci. American 223*, no. 4, 19-29.

Discussion

Hayes I just want to make one point about Professor Beale's comment on the XYY individuals who are branded as psychotics. It seems to me that the ethical point here is not whether or not they should have their genotype inscribed on their passports, but the fact that misconceptions have arisen, and have been spread, about the actual relevance of their XYY genotype to their characters. I think the important ethical point is how did this information get around. It's quite clear that the opinion linking the XYY genotype with higher than average probability of anti-social behaviour must have come from the interpretation of insufficient information. In the future we should very vigorously institute research to establish the significance of suggested correlations before they are disseminated possibly as misconceptions among the general public or the medical profession or penologists.

Beale I entirely agree with that. Of course, it's all very well to say that you shouldn't release information to the press or the public until it's absolutely certain that it is correct, but in practice this is really impossible. Many people obtain information which is provisionally interesting, and important enough to be published, even though it may later turn out to be incorrect. This may not matter with purely scientific reports but it certainly does when there are liable to be serious social consequences.

S. Rose I'm glad you came back, Professor Beale, to the question which Monod raised concerning the problem of the determination of the genetic basis of intelligence. I think this is a point which is of much more immediate relevance perhaps than many of the more distant consequences that we have been considering. In fact here is a case in which certain groups have been publicly branded by the application of quasi-scientific evidence. It's not a question of *stopping* research in this area but rather a question of making absolutely clear that the sort of research that is being done cannot conceivably, on scientific grounds, produce the sorts of conclusions that are being based on it. It is important that we say this very clearly because we are dealing with a particularly sensitive political issue.

Seven Ethics and eugenics
L. S. Penrose
Emeritus Professor of Human Genetics,
University of London
The Kennedy – Grafton Centre,
Harperbury Hospital, St Albans, Herts.

Eugenics is a word invented by Francis Galton in 1883.[1] It was coined to describe a mixture of science and dogma with its aim to improve the human race by judicious marriages of gifted people, on the one hand, and by checking the birth rate of the unfit on the other hand. Thus, natural selection was to be replaced by artificial genetical methods.[2]

Galton himself was a humanitarian but his followers were not always similarly orientated. Proposals for compulsory sterilization of those thought to be carrying unfavourable genes were not uncommon in the early part of this century in England and, particularly, in the United States. Even respectable agitations for voluntary sterilization were tinged with the political suggestion that the lower social groups were biologically inferior and should be discouraged from breeding. There has always been great confusion produced in the minds of the unscientific by the similarity between true biological inheritance and environmental inheritance. It has been seriously argued that poverty is a genetical phenomenon, along with tuberculosis and malnutrition, because it runs in families.[3] In the nineteen-thirties, eugenical ideas got completely out of hand in Germany and they were used to justify all kinds of brutal and criminal procedures for purification of the stock.

One objection to eugenics is its inefficiency. In North Carolina, for instance, where eugenic sterilization laws had been in force for nearly twenty years, it was observed that 71,000 mentally deficient people had been rendered infertile.[4] Very few of them would have had any offspring so the effect of this eugenic action on the population would not have been detectable. At the beginning of the Nazi revolution, instructions were issued to sterilize all inmates of mental deficiency institutions with the exception of those patients whose disabilities could be proved to be environmentally caused. From the eugenic point of view these procedures were quite useless because the populations sterilized were exceptionally infertile. Such laws, however, can persist and become menacing in the hands of unscrupulous politicians and, in Germany, they were the prelude to fantastically horrible massacres, supposedly carried out in the interests of producing a purer Nordic race.

In consequences of the excesses of the Hitler régime, classical eugenic propaganda fell into disrepute. However, in recent years, a new set of ideas has arisen, replacing the old ones, under the name of genetical engineering. The plan is to use modern scientific techniques, borrowed from biology, physiology and genetics, to influence human reproduction. A number of the proposals have nothing to do with heredity in the biological sense as, for example, the transference of an already fertilized ovum from the mother to a different woman or to a test tube. Any effect of the experiment on the offspring would be environmental.

In the early days of eugenics, improvement of the race by fostering the inbreeding of élite groups and discouraging that of the inferior or unfit was favoured. The proposed methods for achieving these ends were crude. The élite races were to be produced by inbreeding between the best stocks rather as prize fancied strains of dogs or horses have been obtained. It is noteworthy, in this connection, that the homozygous fixing of genetical lines by inbreeding is irreversible and that, unless exactly the right kind of animal has been developed, the exercise will have been a failure. Indeed there is much evidence that genes are often more favourable in heterozygous than in homozygous form and that mongrels, who tend to be very heterozygous, are, on the whole, more biologically fit than are pure bred animals. Moreover, genetical variation, which R. A. Fisher called 'the energy of the species', is preserved by mixing breeding.

Negative eugenics attempts to get rid of bad genes from the population. Its methods are to prevent inferior stocks from propagating, to discourage carriers of particular unfavourable genes from having offspring and, if necessary, compulsorily to sterilize, or even kill, those individuals who are thought to be genetically dangerous. In civilizations, where the sanctity of the individual is fundamental, there can be little doubt that all compulsory eugenical procedures are unethical. Many attempts have been made to introduce humane or voluntary systems of eugenic sterilization[5] but, from the point of view of improving the genetical structure of the population, their effects are negligible. Even much more radical proposals founder on two hidden reefs. The first difficulty is that most typical hereditary diseases are recessive and that carriers of the genes that cause them are widely distributed in the normal population. Killing off the abnormal homozygotes has hardly any effect because natural selection will prevent them from breeding anyway. The second difficulty is that the loss of abnormal genes is continually being replenished in the population by fresh mutation. This occurs spontaneously on account of mechanical errors in gene replication and natural radiation to which are added environmental accidents produced by fall out, medical X-ray and virus infections. And, furthermore, not all mutations are necessarily bad and evolution of the human species to its present leading position has depended upon the variations for which they have ultimately been responsible.

Galton's idea was to increase the rate of human genetical change by eugenics in the same directions as that produced by natural selection, only more humanely and efficiently. Genetical engineering, however, has no such clearly specified aims. Let us examine a few ideas. The most practical one is mass artificial insemination. Here the eugenical idea is to produce a very large number of children from one particularly desirable male. There are obvious dangers in this because a dictator, who was powerful but not necessarily genetically desirable, might spread obnoxious genes very widely. The attempt to produce a race of geniuses, by preselected mass insemination, might also fail because, at the present time, we have no precise knowledge of the genetics of genius. It so happens that practically all that is known about the effect of specific genes (DNA) applies to defects, not to assets.

Similar types of thinking have led people to suggest that a way of improving the race will be to duplicate selected individuals by growing identical twins from their cultured cells. This, at the moment, is only a dream of science fiction but is theoretically possible. It would mean that, in the laboratory, identical twins of any desired genius, like Einstein, could be grown, as many as required. They could, of course, be made at different ages from frozen cells and they would need to be nurtured without parents until adult life, at the expense of the state. It is also not proved that a genius in one situation is necessarily a genius in all cir-

Ethics and eugenics 97

cumstances. Moreover, the replicated geniuses might, by accident, be carriers of hidden unfavourable genes.

Somewhat more genetical is the suggestion that it may be possible actually to change genes in the germ cells by substituting better patterns in the DNA than are there already. This kind of substitution is possible in organisms with very simple chromosomal apparatus but it is quite unrealistic to imagine its being made to happen in mammalian germ cells on the basis of present technical genetical knowledge. As far as man is concerned, such engineering would have to be exceedingly exact otherwise bad effects might easily counter-balance the good. Attempts have been made to alter genes in the body by introducing a specified virus; this may be expected to attach itself to the host DNA, and replace an abnormal by a normal gene. The effect of viruses on nuclear DNA, however, has been repeatedly shown to cause chromosome breakages with consequent structural alterations and other dysgenic events.

There is, however, another, quite different, way in which genetical knowledge can be applied in the service of humanity and, at the same time to some extent, produce changes in a eugenic direction. I refer to the medical uses of genetics, which have sometimes been called phenotypical engineering, I believe. A classical example is the use of knowledge of the blood groups to avoid disasters in transfusion. Later came the discovery of the Rhesus factor and the development of measures to counter the effects of maternal and foetal incompatibility. And, at the present time, the study of tissue antigens is helping to prevent rejection of organ grafts.

A quite different application of genetics concerns the detection of conditions which arise on account of hereditary susceptibilities to environmental agencies, especially drugs. By avoiding exposure of genetically susceptible people to particular agents much suffering can be prevented.

An important branch of medical genetics involves counselling prospective parents on the risks that their children may have serious hereditary diseases. A simple example is the increasing likelihood of a child's having Down's syndrome as the mother's age advances. If no children were born to mothers older than thirty-eight years, about one-third of the cases of Down's syndrome would be prevented. Several chromosomal mutations which cause diseases are, in this way, dependent on maternal age but others are not. One of the most important types of chromosomal error is independent of parental age and this can occur if a parent carries an abnormal chromosomal arrangement which has arisen spontaneously, harmless in the parent but liable to cause a disastrous defect in the child. In such cases, information based upon knowledge of families, in which similar errors have been recorded, can be used to warn parents against serious risks. To do this the chromosomes of the parents must be examined but this may not be enough in most cases to give an exact prognosis.

The incidence of some chromosomal mutations can, as we have seen, be influenced by altering parental age. With respect to gene mutations, the only known method of reducing their occurrence is to avoid all but inevitable exposures to radiation.

In recent years, examination of the foetus, at as early a time in development as possible, has become a practical proposition though the techniques in use are still in an experimental stage. Some biochemical errors and chromosomal aberrations are now theoretically detectable at about the fourth month of pregnancy. If a foetus is proved to be abnormal, termination can be advised. The all important ethical principle here would seem to be that the mother, herself, must always be allowed to make the decision in the light of information given by the genetical adviser.

There is one problem which frequently recurs in the minds of people and which concerns the genetical effect of medicine on the human race. It is feared that, by alleviating or perhaps even curing hereditary diseases, carriers of unfavourable genes will be more fertile than before and so will spread these genes,[6] with gradual deterioration of the race. But there are two reasons why we need not be worried by these suggestions. The first is that calculation shows that any increased incidence of a disease like, say, phenylketonuria, which might ensue if all affected subjects became fertile, would be very slight in each generation, one more case in every hundred cases. In conditions where the genetical factors were only a part cause, as in diabetes, rheumatism, and schizophrenia, correcting them can only have an effect on gene frequency in so far as they are hereditary diseases.

The second reason why there is no cause for alarm is that, as soon as a genetical disease can be cured, it ceases to be a genetical risk. I have often pointed out that lack of body hair in man is a genetical disease which we cure by wearing clothes; without them we should all die of cold. This emphasizes that, in one environment, a condition may be disadvantageous and, in another, it might be advantageous and we cannot tell the needs of future civilizations very clearly.[7]

The logical aim of genetical engineering would be to produce a perfect human race, exactly adapted to its environment and, in the distant future, this might mean building men who were specially adapted for living on the moon or in other parts of the universe. Clearly, at the present time, it is impossible to foresee all the needs.

Some enthusiasts, like H. J. Muller, have suggested that the production of some kind of perfect DNA should be the aim of human breeding[8] but, unless DNA takes control and teaches us how to do this, it is unlikely to occur; and the ethics which DNA might teach us would probably be very different from the humanitarian ideas to which we are accustomed.

In conclusion, I would like to mention that, in recent years, the immediate problem of world over-population has, to a considerable extent, eclipsed the more refined technical requirements of eugenics which would, in any case, be concerned with quality rather than quantity of individuals. And it seems that the ethics of population control by some kind of physiological engineering present a far more acute practical question than the ethics of eugenics which I have been discussing.

References

1 Galton, F. (1883). *Inquiries into Human Faculty and its Development*. London, Macmillan.
2 Pearson, K. (1912). 'Eugenics and public health', *J. Roy. San. Inst. 33*, 304.
3 Lidbetter, E. J. (1913). 'Nature and nurture – a study in conditions', *Eugen. Rev. 4*, 54
4 Woodside, M. (1950). *Sterilization in North Carolina*. London, Geoffrey Cumberlege.
5 Blacker, C. P. (1945). *Eugenics in Prospect and Retrospect*. London, Hamish Hamilton Medical.
6 Penrose, L. S. (1950). 'Propagation of the unfit', *Lancet* ii, 425.
7 —— (1963). 'Limitations of eugenics', *Proc. Roy. Inst. 39*, 566.
8 Muller, H. J. (1967). 'What genetic course will man steer?', in J. F. Crow and J. V. Neel (ed.), *Proc. 3rd Int. Cong. Hum. Genet.* Baltimore, Johns Hopkins, 521.

Discussion

Lal Surely the ethics of population control and eugenics are going to become more and more entwined. If, for example, a scheme is adopted so that all parents can have two children, then obviously someone will say: 'But why shouldn't professors be allowed to have more children than labourers?'

Penrose I think there is one thing which should be emphasized. For intelligence (and the same is true of many other important common variations in man) we have no exact information about the positive things, we have only information about the negative things. That is, we know of a great many genes which will damage intelligence and also chromosomal errors which will damage intelligence but we know practically nothing about any gene mutation which will improve intelligence. All the talk about inheritance of intelligence should be very, very guarded. There is no indication that any social group has any different set of genes from any other.

Pontecorvo It seems to me that it is absolutely playing the ostrich to talk as Professor Penrose did; we are just deluding ourselves if we think that human genetic engineering is so in the realm of science fiction that we don't need to start thinking about it. My worry is that the advances will be extremely slow and minor to begin with; for instance, I would estimate that within four or five years it will be possible to cure, to a very minor, limited extent, by genetic engineering, certain genetic deficiencies; nobody will object to that and so we will go on to the next step, and the next step, and so on. And if we don't start discussing these matters now, we shall get to the state, as we did with the atom bomb, when nobody knows what is going on.

Monod Since you insist that this is much closer than other people believe, would you tell us what are the specific points about genetic engineering which you believe raise problems that should be discussed within the scientific community, and also with the public.

Pontecorvo For instance, you get out cells from an

individual who is affected by a deficiency disease and, while these are growing and dividing in tissue culture, you can add some genetic material from normal cells which will result in a normal version of the deficient gene being incorporated into the defective cell. This has already been done for a genetic deficiency which leads to the lack of a particular enzyme. Then you can put these transformed cells back into the same individual, and with a little imagination and science fiction you can see how the transformed cells might express themselves. I am sure that what are relatively minor technical problems in this final stage will be fairly easily overcome. Now this is the first step. It's anybody's guess what the next step will be. I would think that some combination of these genetic transformation techniques with those to be discussed by Dr Edwards for the fertilization of ova *in vitro* is likely. There are innumerable possibilities, and we have to be aware of them.

Eight The obstetrician and genetic investigation

D. V. I. Fairweather
Professor of Obstetrics and Gynaecology,
University College Hospital, London

Much of the work described in this book takes place in the laboratory and relates to plants or animals. The ultimate object of a substantial proportion of the work is ostensibly to benefit the human race and certainly in the medical field this is usually the case. The translation of the findings of non-human research to the human context is a hazardous one and there are many well-known examples of problems which have arisen from the assumption of similarity of action of say, drugs, between species. For instance, one particular drug was found to be a potent ovulation inhibitor in the animal species tested and when introduced originally as a contraceptive agent in the human field, proved extremely embarrassing as it turned out to be quite a potent ovulation stimulator in the human, and is now in fact used as one of the fertility type drugs. The problems and responsibilities of the clinician link-man in such matters must be obvious to all. My brief is to focus on the role of the obstetrician in relation to genetic investigations. With advances in the field of genetics, it has now become possible to obtain certain information about the chromosome and biochemical make-up of the foetus while it is still *in utero* and consequently from this to be able to detect abnormalities in these respects at relatively early stages of pregnancy. To obtain this information it is necessary to sample the liquor amnii, the fluid in which the foetus floats *in utero*, by putting a needle through the maternal abdomen and into the uterine cavity – the technique of amniocentesis. This technique has been safely and beneficially used for many years in pregnancies after the twentieth week to provide liquor samples used to assist in the management of pregnancies complicated by severe rhesus disease. In genetic investigations, however, information is sought at an earlier stage of gestation and amnio-

The obstetrician and genetic investigation 103

centesis between fourteen and twenty weeks of pregnancy is possible.[1] However, experience with amniocentesis before twenty weeks is still rather limited, and may not be entirely without risk. The possibility that the foetus may be damaged or even that abortion may result cannot be discounted, and reference to experimental work with this technique in the animal field suggests that certain types of foetal abnormalities may actually be produced by the amniocentesis.[2] It is with this background that the obstetrician, like myself, is faced with the request to be the plumber to provide the scientist or the geneticist with the necessary material for analysis. Supposing a satisfactory sample of liquor is obtained, techniques for culture of the cells from the sample are still not able to offer 100 per cent success in their growth. Indeed success rates for satisfactory culture vary between 70 and 90 per cent and are in some cases even much lower.[3] Add to this the fact that culture and preparation takes between two and three weeks, meaning that if amniocentesis is performed at say fifteen weeks' gestation, the results of the analysis will not be available until about eighteen weeks. Now consider what is going to be done with the information when it is forthcoming. Supposing the hazards mentioned are overcome.

At the present time if we are appraised of some abnormality we are not yet in a position to correct the abnormality *in utero*, supposing it is chromosomal. The only course of action is therefore to offer termination of the pregnancy if the abnormality is considered to warrant this and here again the obstetrician is faced with making the final decision for it is he who must perform the execution. Already we hear that a certain amount of genetic engineering is possible and before long presumably the obstetrician will be faced with a request to fire the booster rocket to change the course of the chromosome pattern *in utero*, hopefully, to put it back on a normal course. Then to check again at a later stage that all is now normal. It is most important that we all, whether we be scientist, researcher or clinician, realize our responsibilities to the human race. By and large, no one seriously questions our aims and our efforts to benefit humanity. But in present day society and with the tremendous power wielded by the mass media for information and moulding of public opinion, it is vital that when information about our work in whatever field you may choose, is given to the public, that it be given in as an authoritative and factual manner as possible and without sensational claims or statements. We are all enthusiasts about our own interests and work and the temptation to influence others by our enthusiasm may be great, but hardly a day goes past without there being evidence of over-aroused public emotions which bring heartbreak to many in their wake as the true state of affairs dawns slowly through the haze of claims and counter-claims. This state of affairs is particularly regrettable in the medical world. Sometimes the press may be blamed for publishing statements out of context, but much misunderstanding could be avoided, I am sure, if more precise statements were made originally, and if spokesmen avoided sensationalism. If necessary, I would be in favour of issuing a written statement which can be compiled in quiet reflection in preference to off-the-cuff comments. Inevitably, others will be asked to comment on original statements which have been released and here again I would counsel moderation. I speak with some feeling in these matters, as one of those who has to explain, for instance, to the anxious and worried expectant mother who has read or heard that nowadays prenatal diagnosis of sex or chromosome abnormalities *in utero* is possible, why I consider it inappropriate in her particular case to submit her to a test which in itself carries some risks and in which there is less than 100 per cent chance of achieving a firm diagnosis at the end of the day because of vagaries of technique. Unfortunately, her source of information may have given no publicity to the dangers, difficulties, limitations and real application of the investigation in question.

When one considers genetic disorders the risk of any investigative procedure has always to be balanced against the risks of occurrence or recurrence of the condition about which information is sought. The geneticist can provide us with some information about the risk of occurrence of a particular abnormality in a given couple as a starting point. What constitutes an acceptable risk, and who decides the acceptable level? Acceptable to whom? In the end it is, and must be, left to the integrity of each individual in a relationship with patients such as described and which I today represent, to reach a conclusion in conjunction with the patient or with the couple as the case may be, about the desirability and safety for implementation of new techniques, investigations and treatment based on study of all the available factual information. It is for this reason that I agreed to contribute today to highlight this particular responsibility we have in directing or sometimes even cushioning the social impact of modern biology. I have purposely avoided reference to specific genetic problems so as not to cloud the general issues. Often, the role of the go-between linking scientific discovery and practical implementation of scientific knowledge is forgotten. I hope that in this paper I have been able to direct your attention to the problems and responsibilities of this role. If we claim freedom of thought or of action in our own particular aspect of modern biology, remember one of George Bernard Shaw's maxims for revolutionists, 'Liberty means responsibility; that is why most men dread it.'

References

1 Nadler, H. L. and Gerbie, A. B. (1970). 'Role of amniocentesis in intrauterine detection of genetic disorders', *New Engl. J. Med. 282*, 596; Nelson, M. M. and Emery, A. E. H. (1970). 'Amniotic fluid cells, prenatal sex prediction and culture', *Brit. med. J.* i, 523.
2 Poswillo, D. (1968). 'The aetiology and surgery of cleft palate with micrognathia', *Annals Roy. Coll. Surg. Eng. 43*, 61–8.
3 Nelson and Emery, *op. cit.*; Riis, P. and Fuchs, F. (1966), in Moore, K. G. (ed.), *Sex Chromatin and Ante-Natal Sex Diagnosis in the Sex Chromatin*. Philadelphia, Saunders.

Discussion

Parkhouse The advance of contraceptive methods, coupled with the propaganda that they receive, means that in the very foreseeable future most abortions are going to be therapeutic on account of foetal abnormalities rather than because a child would be socially or economically embarrassing. We've heard that there are certain hazards in taking samples of the amniotic fluid and also uncertainties in its analysis. Therefore, if we are only concerned with a child that the parents both want as long as it is normal, would it not be better for the parents to be given the opportunity to reject the child as grossly deformed upon birth than to have to make a decision maybe several months beforehand when there is still the possibility – take the case of rubella – that the child may be perfectly normal? I fail to see the distinction between abortion, particularly when it is carried out as late as seven months as has been reported, and killing the newly-born child. I would say again that this only concerns the situation when a child is born to parents who actually wanted a child.

Humphrey I hesitate as a male to speak about this but I have a feeling that if a woman has actually borne a child it's rather different from getting rid of it before it was born. This is possibly a consideration. However, if you take that into account and the child is one which in a sense revolts her or which makes her full of apprehension as regards what is going to happen in the bringing up of it in relationship to the other children she may have, then I think there may be a perfectly good case.

Kernaghan I should like to say that a lot of people along with me believe that such an action would, quite simply, be murder. Nobody seems to be prepared to consider this.

Young Professor Fairweather, do you want to go on bearing the responsibility for advising parents on these decisions on abortion and also on the undertaking of clinical drug trials in your individual capacity or are we reaching a stage where some sort of formal or informal consultation within the profession, and indeed within science, is required?

Fairweather What you suggest would be an ideal situation but I doubt whether it is a practical possibility to cover all the situations you might meet. It is already being done in certain ways, for example by bodies like the Medical Research Council convening meetings of people who have particular experience and responsibilities in a given field. These 'experts' can then discuss what the risks are and attempt to decide what is reasonable in a particular situation. The suggestion of some body, either national or international, sounds very nice, but it would be difficult to organize practically and many individuals are still going to have to take responsibility.

Doll When you have been able to measure the risk really accurately in arithmetical terms – let us say you have determined that a certain procedure carries a risk to life of one in 100,000 – how do you then communicate that to the patient? You discuss it with them and you give them this figure, but how many women going to a doctor have any impression what that specific risk means? I think people like yourself have the responsibility of translating numbers of that sort into experience in everyday life. If one says to the patient, 'the same as the risk of being run over and killed in the course of a year', this has very much more meaning.

Fairweather Yes. I was somewhat amused that you put the question back to me because I think you particularly have been involved in this business of trying to get over to the public the significance of particular risks such as the pill. Certainly comparisons with the risk of being run over in crossing the street are helpful to the patient.

Nine Aspects of human reproduction
R. G. Edwards
Reader in Physiology, Physiological Laboratory
University of Cambridge

Scientific and medical studies that involve human reproduction inevitably raise public interest and concern. The subject is emotive and personal, the foetus is formed hidden in the oviduct and uterus, and the genetic inheritance of the forthcoming child is established at conception. When conception and abortion are brought into public discussion, and proposals are made to control them, explanations must be full and careful so as to inform the general public and one's own colleagues about the emerging issues. Notice, for example, how the Abortion Act has split the attitudes of gynaecologists called upon to implement it. We are likely to encounter similar divisions among our scientific colleagues and the lay public as we progress with our own studies on human conception.

Investigations carried out mainly by Patrick Steptoe and myself have been aimed at developing clinical and laboratory methods for the fertilization and growth of human eggs in the laboratory. For this kind of work, eggs used for fertilization must be taken from the ovaries, and the various stages of development of the embryos must be identified as they grow in the solutions used for culturing them. Before describing our work, some definition of terms is advisable; the stages of development we will be discussing are shown schematically in Figure 1. The ovary normally produces one egg in each menstrual cycle, but can be stimulated to produce several eggs at a particular time by injecting the patient with the hormones known as gonadotrophins. The hormones also cause ovulation to occur, i.e. the process whereby the eggs are shed from the ovary and enter the oviduct. Fertilization occurs in the oviduct, and the egg is now an embryo. It divides first into two equal cells, then into four, eight and so on, a process that is known as cleavage. During this time, it travels down the oviduct towards the uterus. The embryo enters the uterus when it has sixteen cells or thereabouts, although this is only guesswork since there are no available data on humans to guide us on this point. The embryo continues its development in the uterus, and when it has sixty or so cells, a change occurs in

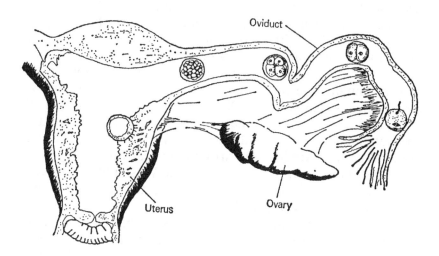

Figure 1 Schematic diagram of the early stages of human development (not drawn to scale). Eggs develop and grow in the ovary under the influence of hormones called gonadotrophins. These hormones also cause ovulation, i.e. release of eggs from the ovary into the oviduct. The egg is fertilized in the oviduct and as the embryo moves down the oviduct it divides into two cells, then into four, eight, sixteen, etc. This process is known as cleavage. The embryo enters the uterus a few days after fertilization, and develops into the stage called the blastocyst. Typically, the blastocyst has a central, fluid-filled cavity lined with cells. The blastocyst then becomes closely attached to the uterus. This is known as implantation.

its structure. Spaces filled with fluid appear between the cells, and join together to form a large cavity which ultimately occupies most of the volume of the embryo. At this stage, the embryo is called a blastocyst. The blastocyst stage is reached just before the embryo becomes attached to the uterus (womb) (see Figure 1). Attachment is called implantation; after implantation, the blastocyst grows rapidly and starts to develop into a recognizable foetus.

We can now recover eggs from women some two or three hours before ovulation, fertilize them in culture using ejaculated spermatozoa from the husband, and grow some of them into blastocysts.[1] This means that all of the stages of human development up to implantation are now open to study. Implantation itself, and the earliest stages of subsequent foetal growth, will perhaps soon be investigated outside the body. Bringing such embryos into the light of day provokes various thoughts and opinions: the beginning of test-tube babies, armies of carefully-planned robots, playing at God in the laboratory. How truthful and meaningful are these ideas? What are the advantages and disadvantages that will actually accrue from this work? In this brief article I will try to outline answers to some of these questions and comment on some of the responses to our work that have arisen already or are likely to arise in the near future. My lawyer colleague David Sharpe from the George Washington University and I have already considered some of the legal and social issues that are likely to arise.[2]

The primary incentive for our work stems from a fundamentally humanitarian view. We are attempting to improve the quality of human life by regulating fertility and by identifying and averting the causes of certain forms of anomalous development. We hope to make it possible for some infertile parents to have their own children wherever we believe our help to be valid and meaningful. We have already studied the causes of some inherited defects, and we will do our utmost to avert or modify the distressing effects of such inheritance. A second stimulus to our work is intellectual: we wish to acquire more knowledge about our own species and its environment. We believe these objectives to be worthy and humane, indeed that it is difficult to find a subject more responsible and challenging than that of improving the quality of human life. Let me first outline the opportunities, clinical and otherwise, that our work can bring, and then turn to a discussion of our responsibilities as we see them.

The first benefit of our work should be to some infertile couples; they may have the opportunity of having their own children where it is difficult or impossible to have them at present. One group of patients consists of those where the wife has blocked oviducts. Transferring their own embryos, fertilized and grown in the laboratory, into the uterus, offers them a real chance of having their own children. Other infertile couples may benefit later, for example those where husband and wife are clinically normal yet fail to conceive, and those where the husband has developed antibodies against his own spermatozoa, or has too few spermatozoa for normal conception.

In the longer term, our work might benefit those couples at risk of having children with inherited defects. Disorders such as haemophilia or a form of muscular dystrophy occur almost exclusively in males, and could be avoided by ensuring the birth of daughters in families which have been identified by biochemical tests or through the birth of one afflicted child. Rabbit blastocysts have already been sexed with complete success by Richard Gardner and myself,[3] similar methods are being applied to mouse blastocysts,[4] and human blastocysts could probably also be sexed by these methods. Improvements in technique are urgently needed, but at least we can now begin to consider such methods for tackling the problem. Sexing spermatozoa would be a better method of controlling sex ratios, but this method does not appear to be feasible at the moment. Other diseases will present greater difficulties. Mongolism is an example; it arises through a chromosomal error during cell divisions, and the causes remain to be identified. The early cell divisions in eggs and cleaving embryos can now be examined, so we should be in a position to analyse the causes of this congenital defect. Whether or not our work will help to alleviate this disorder remains to be seen. The causes of other distressing disorders might also be analysed using our methods. One such disorder is spina bifida, many people with this complaint being permanent invalids. The origins of this disorder are obscure[5]: it might even arise through factors in the cytoplasm of the egg or anomalous conditions in the uterus.

A radically different approach to the amelioration of genetic disorders could arise from methods developed by our colleague, Richard Gardner. He showed that in mice the injection of one or a few donor cells into a blastocyst could modify the characteristics of the resulting offspring.[6] The donor cells are taken from other embryos, and characteristics of the other embryo are partially conferred on the host embryo as it grows. Once again, a great deal has to be learned in animals about this approach, but it might permit, one day, the modification of human genetic disorders that are incurable and crippling. Here, we will have to balance the implications of these measures against the grief and hardships that arise once a malformed baby is born. Many scientists are now deeply concerned about the demands of society made by the increasing proportion of genetically handicapped people surviving in our community.[7] Measures might have to be taken to ease this situation, if possible by eliminating the causes or effects of the birth defects.

Another benefit of our work should be a deeper understanding of the processes of conception, leading perhaps to the development of new methods of contraception. We know so little about human ovulation, for example, as shown by our ignorance until recently of even the time when it occurred. We know next to nothing about fertilization, cleavage and implantation. Yet these are among the very processes that must be studied in the search for safe and acceptable contraceptives.

Two further potential developments of our work will need careful consideration before they are undertaken. One is the continued growth of embryos in culture after implantation, and the initial stages of this study are now feasible. The first steps in this work would be largely for the purposes of scientific enquiry, although the control of implantation would be of potential value in contraception. Many difficulties will probably arise before even the early post-implantation embryos are grown successfully, but this type of work will undoubtedly be possible one day. Foetuses might even be grown to full term in culture, although not for a long, long time ahead. Growth of embryos even through the early post-implantation stages will raise some ethical problems, because the foetuses will almost certainly undergo considerable growth and differentiation of various organs. The second development will raise issues of greater importance. Methods are being developed in animals for transferring the nuclei of cells taken from adults into eggs that were previously enucleated. The embryos so treated might develop largely as a genetic copy of the donor, and this method of initiating a known type of embryonic development is often referred to as 'cloning'. The issues arising from the application of these methods to men have been debated widely in the past, and we have made our own comments.[8] I will not discuss further either the continued culture of foetuses after implantation, or the consequences of cloning, since other issues warrant more immediate discussion.

Great scientific and medical opportunities, together with issues needing close and careful attention, thus accompany the development of our work. For the scientist working with human material, various responsibilities had to be recognized and faced from the outset, and included medical commitments when the work was extended to the treatment of patients. Although many of these responsibilities are owed by the physician to his patients, I believe that the scientist handling the embryos must be judged equally responsible. Much of what I now write has been stimulated by discussion with Patrick Steptoe.

Let me describe briefly the responsibilities that arise in our work. Our primary responsibility at present is to the patients themselves. We have tried to avoid raising their hopes excessively during our initial work. When we first considered asking patients to help us, the early stages of fertilization had been accomplished in the laboratory by ourselves and our colleague B. D. Bavister, using eggs taken from ovaries excised for clinical reasons unconnected with our studies. The clinical application of our work demanded that eggs be removed just before ovulation directly from the patients. They had to be given hormones – four injections over 8–10 days – followed by a minor operation (known as laparoscopy) in order to remove the eggs. The husband had to provide an ejaculate. In the early days there was still a great deal to accomplish before we could even be certain of obtaining embryos. Our decision to ask patients to help was a critical one, and was based on our conviction that research in our field had reached a point where its social application afforded major advantages to the patients themselves, and later to others. The difficulties, demands and opportunities were explained to each patient, although there are obvious problems associated with obtaining what is known as the 'informed consent' of the patients. How can they be expected to understand both the medical and scientific problems when they are probably untrained in both disciplines? Obviously, full understanding is impossible, but I believe they understood enough to know of the demands we would make upon them. Moreover, our patients have included doctors and nurses, and their acceptance of our methods is perhaps sufficient comment in itself. These remarks apply, of course, solely to our rapport with the patient; I am not dealing here with the strictures that apply to the performance of responsible clinical and scientific work – our published papers are the best guide for that.

Now that we have grown embryos through their early stages of development, our next responsibility will be to the foetuses that grow in the mother after the embryo has been placed in the mother's uterus. For the first time in human history we are deliberately initiating embryonic development outside the mother, with the intention of her eventually bringing the baby to full term. We must do all we can to ensure that these babies are normal. It is well known that abnormal children are born under circumstances that are evidently quite normal. Inherited anomalies in babies are largely written off as being outside human control; there is hardly any condemnation of parents who bring more than one deformed child into the world. In our work, fertilization is at random in the sense that the eggs and spermatozoa are unselected, and the risk of inherent anomalies will be as high as in normal conception. But additional errors could arise from accidental shortcomings in our methods, or from the exposure of the embryo to our laboratory conditions. A great deal of knowledge has accumulated to show that animal embryos of several species develop normally when transferred into the uterus of a female. Nevertheless, although these observations are encouraging, their relevance to the human situation remains a matter for debate.

At present it would be unjustified to transfer the human embryos we have grown, until we have shown that their chromosomal constitution is normal. We will use some blastocysts for chromosomal analysis to ensure as far as we can that future babies will be normal. We accept the possibility that some children with abnormalities such as Mongolism may be born from transferred embryos despite all our precautions, and probably in higher frequency than in the general population. This is because our patients are mostly over thirty, an age when mothers carry an increasing risk of having such children. A hypothesis developed jointly with Dr S. A. Henderson concerning the causes of Mongolism would indeed suggest that there is no hope of averting the origin of such embryos since the egg itself might be inherently abnormal.[9] We hope that an alternative hypothesis proposed by German[10] is correct, and that the disorder arises when there is a long delay before eggs are fertilized. We can probably control this aspect of fertilization under laboratory conditions.

Some of the foetuses will be examined to ensure that they are as fully normal as possible. A problem could arise with abnormal foetuses, for we will have to consider the abortion of those with a particular inheritance or with various deformities. This consideration does not apply only to our work, and has been widely debated previously. Such abortions are being carried out at present, after natural conception in families with a known genetic risk. Abnormal foetuses can often be detected using a method known as amniocentesis to obtain a few cells of the foetus for testing. Should the situation arise, we will explain in detail to the patients, and help them to reach their own decision in deciding on an abortion or otherwise. Many doctors already have to cope with this problem. Facing such a decision will be no light matter for our patients who have gone to great lengths to establish the pregnancy. A further problem might arise in knowing where to draw the line on aborting foetuses with inherited or other anomalies. Some people might fully accept a Mongol baby. Sometimes the effects of inherited genes are still ill defined. Hasty decisions could lead to unexpected social consequences; an example arose recently when certain men were found to possess two Y chromosomes. Adverse publicity arose about certain behavioural aspects of this condition, and unjustly influenced the attitudes of society to these adults.

Various ethical issues also arise from our work, together with legal points that will vary widely between different countries. Let me consider first our own attitudes, and then comment on those of other individuals and organizations.

We believe it essential that doctors and scientists are free to pursue research into aspects of knowledge that could contribute to the well-being of humanity provided the rights of the patients, including those of the foetus, are safeguarded as far as possible. The pursuit of knowledge for its own sake is not sufficient reason for enquiry unless tempered by clinical responsibility. Viewed in this way there is no clash between scientific and medical ethics. The absence of children can lead to extreme unhappiness and even to the breakdown of marriage. Our responsibilities and beliefs could fully cover sexing embryos to avert the birth of children with inherited defects. At present, we would be less convinced about choosing boys and girls merely to gratify parental desires, because there would be no controls to maintain equal numbers of the two sexes, and because the great amount of effort needed to achieve this choice would not be justified. Moreover, more discussion is still needed about choosing the sex of offspring to satisfy parental desires.[11] We would have no objection to injecting cells into blastocysts if convincing evidence existed that the health of foetuses was substantially improved by this procedure. But there are scientific problems to be debated first about this method. The donated cells might carry their own deleterious genes.[12] Covering the effects of abnormal inheritance by donor cells might preclude the identification of the offspring with such inheritance in the next generation. We would accidentally create those conditions we are trying to avoid, namely increasing the frequency of unwanted genes in the population.[13] Nevertheless, problems such as these are probably manageable. Our work could thus avert the birth of children with genetic disease or modify the effects of these genes. Such ideas stand in contrast with current methods of attempting to treat such disease after birth; methods which are resulting in an ever-increasing number of patients for treatment. It would surely be far better to select against afflicted blastocysts (if those carrying the mutants could be identified), or modify them, than to abort affected foetuses, or to provide severely handicapped offspring with continual treatment and then advise them not to have their own children.

Our work will almost certainly bring us into conflict with established social attitudes of various sections of the community. Some groups believe that all stages of human life must be protected, and consider that the fertilized egg as a potential human being must be given identical rights to an adult. I find this 'absolute' argument difficult to accept. Unfertilized eggs and spermatozoa are also living entities, and they can be treated in various ways to result in abnormal foetal development; must they also share in this general protection? Mouse eggs, for example, can be stimulated to undergo development without fertilization to advanced foetal stages.[14] The moment when 'life' begins is almost impossible to define; embryonic development is a gradually evolving process. Nor is ours the only field of study where this problem arises. The use of contraceptives such as the intra-uterine device (IUD) which probably expels or destroys blastocysts[15] is accepted because the needs of the couple are considered to outweigh the rights of the blastocysts. Another reason for accepting the IUD is that population pressures are such that the rights of blastocysts must be subordinated to the general good of society. Growing human embryos in culture to the blastocyst adds one more parameter – we have expressly interfered to achieve fertilization. Otherwise, the situations are closely parallel.

I find that it is almost impossible to draw a hard line defining the limits of legitimate interference with foetuses. Different limits will be decided by various societies based on the conditions prevailing in each society at a particular time. Some communities might decide to protect the fertilized eggs (has this ever been true of any society?); others have gone to the other extreme and accepted infanticide. Implantation is assumed by some authorities to represent the beginning of life, since further development is impossible unless the embryo attaches to the mother. I would question such an assumption just as much as that involving fertilization. My view is based on a knowledge of developmental biology: a blastocyst is still a very simple structure and can have few rights, even though a potential individual, as compared with a mid-term foetus, and the latter (in turn) as compared with a full-term baby. This view has been literally accepted in the reform of the abortion law and in the selective abortion of foetuses with inherited abnormalities following their identification by amniocentesis. Infanticide is not acceptable today, but abortion halfway through pregnancy is. Permission to interfere with or destroy the foetus has been withdrawn to earlier stages of pregnancy to gain acceptance. This is a further reason for identifying traits in blastocysts: the ethical problems are reduced because the embryo is in such an early stage of development. Another reason is the considerable waste of human resources and the amount of human misery involved in aborting women in mid-pregnancy. A difficulty with formulating rules about the rights of foetuses of particular ages is that strictures can become outdated almost before they have time to become operative. For example, the beginning of life was once taken as the time when the foetus can first exist independently of the mother, but the definition must constantly be revised backwards with improvements in the care of premature babies.

The rights of the foetus have already been limited in many countries, and doctors in the UK work within the framework of the 1967 Abortion Act.[16] The Church is divided on the issue. It is of interest here to refer to the Church Assembly discussion on abortion[17] that was issued before the Act was passed. The report places responsibility for the termination of pregnancy under some circumstances largely on the doctor. Since the report was written the methods of identifying abnormal foetuses have greatly improved, and so strengthened the doctor's diagnostic powers. Indeed, this and other discussions reveal a great deal about the attitudes of various Churches to the rights of the foetus. Let me quote one passage (p. 24) of this report: 'So with abortion: the primary and general intention of the law has been to preserve as inviolable the right of the unborn child to live; yet the number and extent of the exceptions and accommodations are such that this right to live cannot be described as absolute and *in all circumstances* inviolable' (their italics). The basis of the ethical stance of the Christian Church was founded many years ago, one of its rules being the importance of baptism before death.[18] At one time, it was believed that the adult form was already present in the gametes; new knowledge therefore challenged the earlier concepts. Ethics based on a scientific knowledge of evolution[19] offer me a much firmer basis for debating the values of human life than discussions about the history of religion.

Will our work bring us into conflict with other ideals or concepts of society? The babies we produce will be wanted babies – often desperately wanted. Curiously, the problems of infertility are often belittled at the same time that the advantages of parenthood are stressed. Some people involved in population control have criticized us because of the coming birth of extra children when limiting births is all-important. I find this attitude perplexing, invalid and short-sighted. The infertile cannot be penalized for the sake of the over-fertile, surely we should do all we can to produce children for responsible parents. Our work could also encourage the use of methods of contraception such as tubal ligation ('tube-tying') in women, an operation difficult to reverse. At present many women prefer not to use tubal ligation as a contraceptive measure because they fear permanent infertility in case they should remarry or lose their existing children. Embryo transfer would avert this finality and probably make tubal ligation more attractive to many women.

In carrying out our work, we have met problems similar to those faced by pioneers in other fields. The doctor inducing an abortion in the face of repressive legislation, the early leaders of the Family Planning Association, the first to graft a human kidney, all raised challenges to public values and private commitment. In many of these examples, the first steps were taken by individuals convinced of the correctness of their own attitudes. Individuals and society both might benefit if authoritative assistance could be given in reaching such decisions at critical times, but advice would have to be as broadminded and comprehensive as possible. We have suggested elsewhere[20] that a need might exist for an organization to provide such informed opinion and so assist in decision-making.

In the final analysis, however, and certainly at present, responsibility for new work seems to reside with the individual, based on his full knowledge of the new opportunities. He must then show his decision to have been correct. The intellectual and social rewards of studies on human beings are very great – in my opinion far greater than those obtained by work done solely in the laboratory. But every stage of clinical research must be questioned, and care taken to ensure that it is related to social needs and is not stimulated purely by its own impetus. Scientific and medical responsibility is not a new phenomenon, although different disciplines raise their own novel problems at particular times. I believe that problems arising through studies in or the control of human reproduction raise the major ethical and social issues that face us today. Conflict with other beliefs, or responsibility to patients, arises less frequently in science than in medicine. When such issues do arise, we must clarify our own attitudes, and persuade others that our own decisions are correct.

Acknowledgments

I wish to thank Patrick Steptoe especially, and also Professor C. R. Austin and my wife, Dr Ruth Fowler, for our various discussions on topics covered in this paper.

References

1. Steptoe, P. C., Edwards, R. G. and Purdy, J. M. (1971). *Nature* 229, 132.
2. Edwards, R. G. and Sharpe, D. J. (1971). *Nature* (in press).
3. Gardner, R. L. and Edwards, R. G. (1968). *Nature 218*, 346.
4. Gardner, R. L. (1970), 'Intrinsic and extrinsic factors in early mammalian development', in Raspé, G. (ed.), *Advances in the Biosciences*, 6 (in press).
5. Carter, C. O. (1969). *J. biosoc. Sci. 1*, 71; Nance, W. E. (1969). *Nature 224*, 373.
6. Gardner, R. L. (1968). *Nature 220*, 596.
7. A recent symposium was devoted to this topic: Conference on Genetic Disease Control, Washington, December 3–4 1970. Sponsored by the US National Cystic Fibrosis Research Foundation, the US National Genetics Foundation and the US National Institute of General Medical Sciences.
8. Edwards and Sharpe, *op. cit.*
9. Henderson, S. A. and Edwards, R. G. (1968). *Nature 218*, 22.
10. German, J. (1968). *Nature 217*, 516.
11. Edwards and Sharpe, *op. cit.*
12. See the paper by Mintz, B. (1970) in *Genetic Concepts in Neoplasia*, Baltimore, Williams & Wilkins, 477. The effects of mutants in donor and host tissues on the characteristics of

chimaeric mice are described in this paper. Mintz also has a comment on this topic following the paper by R. G. Edwards, given at the Harold C. Mack Symposium on the Biology of Fertilization and Implantation, Detroit, October 1970.
13 See the comment by L. S. Penrose (p. 98) on the consequences of phenylketonuriacs having their own children. He estimates an increase of only 1 per cent in the number of phenylketonuriacs born per century. This comment is relevant to my discussion on the cure of genetic disease.
14 Mouse parthenogenones can develop to advanced stages of foetal development. Tarkowski, A. K., Witkowska, A. and Nowicka, J. (1970). *Nature 226*, 162.
15 Eckstein, P. (1970). *Brit. med. Bull. 26*, 52.
16 See the Memoranda on the Abortion Act, 1967, and on the Abortion Regulations, 1968, issued by the Medical Defence Union, 1968.
17 Discussions by the Church Assembly Board for Social Responsibility, issued by the Church Information Office, Church House, Westminster, London, S.W.1. Various reports have been issued, including one entitled *Abortion, An Ethical Discussion*, issued in 1965, reprinted 1968.
18 Discussed fully in the book by Williams, G. (1958). *The Sanctity of Life and the Criminal Law*. London, Faber & Faber.
19 A pertinent comment is found in the essay by Huxley, J. 'The New Divinity' in *Essays of a Humanist*, first published by Chatto & Windus, 1964, then as a Pelican Book issued in 1966, and reprinted in 1969. Many other essays in this book are highly pertinent to this topic.
20 Edwards and Sharpe, *op. cit.* (see n. 2).

Discussion

Silverstone Everybody of course is conditioned by the ethical and belief systems of the society in which they live. For many women, having a child is the *raison d'être* that they are offered by society rather than becoming a big shot biologist. So I'd like to know if you discuss these fundamental concepts when you offer these people the facts to make their 'free of external pressure' choice?

Edwards I believe that no decision that we take anywhere, at any time, is free of external pressure. With patients of

Aspects of human reproduction

different religious convictions, for instance, how can we advise them equally, how can we possibly know that they take this decision free of external pressure? What we can do is to explain clearly the problems that would arise with a particular foetus, give them any other relevant advice, and leave the decision largely to them.

Singer It's actually possible, as Professor Fairweather knows, to take just a small amount of amniotic fluid at between sixteen and twenty weeks to do nuclear sexing with, if I recall correctly, 98 per cent accuracy in approximately 300 cases. So by choosing on the basis of this test whether to abort or not, one could choose the sex of babies which are born.

Edwards I consider that suggestion to be unjustified. My general attitude to selective abortion is as follows. Certain foetuses are so malformed that I would have no hesitation about selective abortion. I cannot see any reason for keeping those alive and I would probably take this same attitude at birth. This is what happens now in that people take a decision not to sustain the new child or attempt to keep it going.

Chater What happens when you are unable to obtain a good egg from a woman who wishes to have a child? Would you then consider obtaining an egg from another woman, and, if so, how would you choose the donor?

Edwards This will raise tricky questions; using an egg from another woman would be like an AID in reverse. Should we ask a sister to provide an egg to keep the offspring within the family, so to speak, or should we select a complete and total stranger to avoid any psychological relationship between the egg donor and the child? I don't know, but one point is worth making: AID with a male donor is widely accepted, even, I believe, by some parts of the Christian Church. If we used egg transfer, the baby would belong more to its parents than one conceived by AID because the mother would be the natural mother in one sense – she would be the uterine foster-mother – and the father would be the true father.

Part four Immunology and cancer

Ten Some implications of modern immunology

J. H. Humphrey
Head of Division of Immunology and Deputy Director,
National Institute of Medical Research
Mill Hill, London

Although immunology has now branched out in many directions, it is important to remember its origin as that part of bacteriology concerned with studying how man and other animals become immune to infectious diseases – or, more generally, how animals have maintained their integrity and evolved in environments which contained plenty of microbes (bacteria, viruses, protozoa, etc.) capable of colonizing and destroying their hosts. This may appear to be a limited, even though a very important field. However, the more we have learned about immunity and the immune response the more subtle, adaptable and complex have proved to be the mechanisms involved. I do not propose to go into them in any detail, but rather to discuss very briefly how immunology and molecular biology have impinged upon one another and to say something about the understanding of the immune response which has emerged. With this understanding, incomplete though it is, has come the possibility of manipulating the immune response, and the capacity to do things which have had, or can have, socially important consequences.

Put very shortly, and grossly oversimplified, the main protection against microbes which have succeeded in penetrating the body's rather effective and self-repairing mechanical barriers lies either in their being killed at once by substances in the body fluids, or, more usually, in their prompt ingestion and subsequent digestion by the ubiquitous specialized scavenging cells, the macrophages and the microphages whose function was first recognized by Metchnikoff some eighty years ago. Microbes which are not killed and are able to multiply within the body cause diseases, especially when they also manufacture potent toxins. Recovery from microbial diseases involves eventual elimination of the microbes by essentially similar means, which have become much more efficient during the course of the disease – as they must have done, since there are many more microbes to eliminate. Such defence mechanisms must be able to recognize microbes or their products as 'foreign' – that is they must distinguish between 'self' and 'not-self'. Furthermore, since microbes come in many guises, each chemically different, and since there are no structures characteristic of all microbes, the range of 'foreign' molecules which can be recognized has needed to be very large indeed and to include practically any large molecule whether or not of microbial origin. How this recognition works is the main theoretical question in immunology.

In principle we know the answer – the main recognition mechanism, though there could be others less specific and of lesser importance, is by means of antibodies present in the blood and other body fluids. Antibodies belong to a family of broadly similar proteins collectively known as immunoglobulins, and their characteristic property is to have combining sites which can specifically interact with quite small surface configurations on foreign molecules. As already mentioned, these do not have to be microbes or their products. They can, for example, come from pollens, moulds, or dusts which penetrate the body after being breathed in through the lungs; or they can be materials introduced by an abrasion or an insect bite, or by a syringe and needle – but invading microbes are the commonest source. When antibodies combine with toxins they neutralize their toxicity, and when they combine with microbes they coat them so as to prevent them multiplying and to make them much more susceptible to uptake and killing by the scavenging cells. Within the family of immunoglobulins there are a number of different main classes containing molecules easily distinguishable on the basis of their size, shape and chemical composition, and within the classes are subclasses distinguishable by more subtle differences. Immunoglobulins belonging to different classes, although they may react with the same foreign molecules, have different biological properties. For example, some but not all pass from the mother to her infant; some activate important auxiliary mechanisms collectively known as 'complement' which are responsible for many of the manifestations of inflammation and also kill microbes; one is preferentially concentrated in external secretions, and has been likened to an 'antiseptic paint'; another is responsible for the manifestations of allergies such as hay fever, urticaria and some forms of asthma. Similar classes are found in all higher vertebrates examined. Presumably they have evolved because each has conferred some survival value during the course of evolution.

All immunoglobulins have a similar basic structure. They are composed of two different kinds of polypeptide chains, a smaller one termed 'light' (or L) and one about three times as large termed 'heavy' (or H). These are linked together in various ways in different classes of immunoglobulin so as to produce molecules of different shapes and sizes but all contain the general pattern

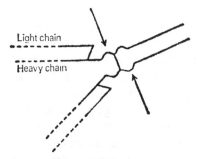

Figure 1 Diagrammatic representation of the four-chain structure of IgG, in which the chains are linked by three disulphide bridges. The broken lines indicate stretches of the chains in which heterogeneity occurs within a given chain type; these stretches include the antibody-combining sites. The undulating portion of the heavy chain represents a portion susceptible to cleavage by proteolytic enzymes (papain at the point marked by the upper arrow and pepsin at the point marked by the lower arrow).

shown in Figure 1. This is a Y shape, with flexible arms which carry at their ends the 'combining sites', whose structure is adapted to combine with some patches on the surface of a foreign molecule. During the past ten years since this structure was first proposed by R. R. Porter[1] we have learned a great deal about these quite remarkable molecules. On the one hand they have been recognized as being extremely well designed for the functions which they fulfil, but on the other hand, their origin and genetic control do not fit easily into the schemes which have been found applicable to other proteins, and may require some revision or extension of current ideas on genetic control of protein synthesis.

Each polypeptide chain (both the L and H) of any given immunoglobulin consists of a part whose amino acid sequence is constant from molecule to molecule, as is the case for all usual proteins. Where there are hereditary variations in the constant part these obey the ordinary rules of genetics, and this part of each chain appears to be controlled by quite a small number of genes, suitably termed C genes. These determine those properties of the molecules on which depend their various different biological functions, some of which were mentioned above. What is unique about immunoglobulins, however, is that the order of the amino acids in the other part of the L and H chains varies from molecule to molecule, so that at the so-called 'variable' end no two molecules are likely to have the same shape. It is the variable parts of the L and H chains which contain sites which have shapes complementary to (and so able to bind firmly with) some patch on the surface of foreign molecules, and confer on an immunoglobulin molecule the property of being a specific antibody. There is very strong evidence that the variable parts of each chain are controlled by separate V genes, although the whole chains are synthesized as single units. The number of different amino acid sequences in the variable region of the immunoglobulins made by any normal individual is certainly very large – I would guess at the very least 100,000 – and this implies the existence of a corresponding number of V genes. To explain how this enormous diversity of genetic information for the control of part of one kind of molecule comes about faces both immunologists and molecular biologists with a major problem. Explanations in terms of current genetic ideas, as outlined by Professor Hayes, would require a very much higher mutation rate for the V gene in the cells which make immunoglobulins than is known to occur for any other genes in either microbes or higher animals. Various hypotheses have been proposed to explain how this might come about, of which the most comprehensive is due to Gally and Edelman.[2] This suggests that there are quite a small number of structural genes (perhaps 100–200) of which parts can break away and readily recombine elsewhere within the chromosome in which they are situated, thus altering the order of the nucleotides and consequently the amino acid sequence in the peptide chains coded by them.

The hypothesis is based on the well-recognized fact that parts of bacterial chromosomes can become split off as separate functional units known as 'episomes', but it would certainly require the existence of some very special mechanism for ensuring that frequent and ordered translocation of parts of neighbouring genes should occur. Whether there is such a mechanism remains to be discovered.

A millilitre of human blood contains about 2×10^{16} molecules of immunoglobulin, most of which – as I have explained – differ from one another. They include all sorts of recognizably useful specific antibodies as well as others which may be useful but whose specificity and function are not known. Like other proteins in the blood, antibodies are continuously being removed and broken down, and replaced by new molecules; about half disappear and are renewed every three weeks or so. In order to understand the immune response it is necessary to consider briefly what cells make them and how they are stimulated to do so. The cells responsible are lymphocytes, of which an average adult possesses some 10^{12}, but whose function in the immune response has only become clear in the last ten years. They are rather nondescript, undifferentiated cells, consisting of little more than a nucleus (which contains, of course, all the genetic information common to all other cells of the body) and enough cytoplasm to keep the cell alive. However they are able, when suitably stimulated, to multiply and some can also turn into very active cells capable of making large amounts of immunoglobulin. They have a complicated life cycle, which is still not fully understood. From the time of birth onwards stem cells, the precursors of lymphocytes, are formed in the bone marrow, and released into the blood stream. Some go straight to 'lymphoid' tissues (spleen, lymph nodes, the tonsils, appendix, etc.), after which they live an independent existence, spending part of their time in these tissues and part circulating in the blood stream. They are subject to continuous death and replacement by new cells which are formed when some of them are stimulated to divide – for example by contact with a specific antigen as described below – but how frequently this happens is not yet certain. These cells which arise in the bone marrow and colonize lymphoid tissues are those which, when suitably stimulated, can differentiate into antibody secreting cells. Other stem cells from the bone marrow go to a special lymphoid tissue, the thymus. There they turn into lymphocytes which divide continuously. Most of these die within the thymus, but a proportion escape and in turn colonize the other lymphoid tissues, where they continue an independent existence much like that which was already described. They differ permanently, however, in several ways: they occupy distinct parts of the lymphoid tissues; they live for a very long time (up to several years) unless stimulated to divide; they do not secrete antibodies or turn into antibody secreting cells; they are nevertheless able to interact with antigens and, if these are on other cells, they are able to kill them; after interacting with antigens they stimulate macrophages to ingest and kill microbes more efficiently; and in a rather complicated way they help antigens to stimulate other lymphocytes to secrete antibodies.

There are strong grounds for believing that lymphocytes bear immunoglobulin molecules at their surface, even when these molecules are not actually secreted. Furthermore it appears that each lymphocyte bears only a single kind of immunoglobulin, and that the great majority of lymphocytes each bears a different immunoglobulin. The inference is that by the time lymphocytes have settled in the lymphoid tissues mutation of their immunoglobulin genes has occurred (perhaps by some sort of mechanism such as that proposed by Gally and Edelman) and that the total population is made up of cells each able to make only one kind of immunoglobulin, but between them representing the enormous variety which we know to be present in the blood. It is a plausible but unproven hypothesis that the acquisition of this genetic diversity occurs in the course of the continuous cell division which the lymphocyte population is known to undergo.

The immunoglobulins at the surface of lymphocytes serve as receptors for antigens. If the antigen is able, by chance, to bind strongly with the immunoglobulin on a particular lymphocyte it can affect that lymphocyte in one of three ways:

1. It may, paradoxically, inactivate or kill it. The consequence of all the cells capable of reacting with a particular antigen being put out of action is known as immunological tolerance or paralysis, since the animal now lacks the means of recognizing the antigen as foreign. It is very important that this should be a possible result, since otherwise during the course of development lymphocytes would arise able to react with the body's own constituents, with disastrous consequences. When, for various reasons, the mechanism fails at later stages of development autoimmune disease occurs. By the same token, the mechanism can be exploited, for example to make an individual tolerant of the foreign antigens on another's tissues for the purpose of transplantation.

2. It may stimulate the lymphocyte to differentiate into a cell

which secretes large amounts (thousands of molecules a second) of the same kind of immunoglobulin as was present at the surface – i.e. to make a substantial excess of antibody specific for the stimulating antigen.

3. It may cause the lymphocyte to multiply, thereby producing a large number of identical daughter cells bearing the same immunoglobulin and capable of being stimulated by the same antigen as the parent lymphocyte. In this way the proportion of such cells in the total population is increased, and a correspondingly bigger and more rapid immune response results when the antigen is reintroduced into the body. This is the basis of what is often termed 'immunological memory', and accounts for the lasting immunity which follows many infections (mumps or measles, for example) and successful prophylactic immunization. The proportion of lymphocytes which are stimulated by the antigen to respond in one or other of these ways depends upon the nature of the antigen, on how much is present and for how long, and on factors in the local environment in the tissues in which they meet – which have aptly been termed by Medawar 'the proper diplomatic channels', though we do not know too much about them.

To recapitulate very briefly, an animal possesses a built-in capacity to make antibodies of very many specificities, and this capacity is represented by a large population of cells each of which bears receptors corresponding to the sort of immunoglobulin which it can make. The receptors interact with chemically defined structures, irrespective of whether these are present on microbes, on adventitious materials (such as dusts or pollens) introduced into the body through the air passages and lungs, or on molecules injected with a syringe. Immune responses are the consequences of such interactions, and are due to the selective stimulation and magnification of a pre-existing capacity. Their control involves selectively increasing or decreasing the particular cell population which represents this capacity, and it is not a question of imposing entirely new responses. I do not want to rule out the possibility of introducing artificially the genetic information to produce a new kind of antibody into lymphocytes cultured *in vitro*, but at present for practical purposes it is easier to select from the vast amount of genetic information already available in an individual's lymphocyte population.

Since the purpose of this book is not simply to summarize various aspects of modern biology but to consider their implications, it is time to discuss some of the social consequences of our ability to manipulate immune responses fairly effectively, even if not exactly at will. I will first discuss situations in which immunity can be increased; of which by far the most important is prophylactic ('artificial') immunization.

Prophylactic immunization

Living microbes, or their various toxic products, are usually powerful antigens and, when present in the body, rapidly stimulate antibodies to be made against them. This occurs both when microbial infection causes a full-blown disease – recovery from which depends upon mobilizing the body's defences more rapidly than the microbes can multiply – and when the infection is so mild as to be almost inapparent. Many individuals possess a degree of immunity to the potentially infective agents prevalent in their community as a result of inapparent infections, but immunity breaks down when the infective dose is very large, or when an unfamiliar organism is introduced (e.g. in epidemics of plague, smallpox or influenza). The great practical achievement of the pioneer immunologists at the end of the last century was to recognize that effective immunity could be stimulated by killed microbes, or by their toxins whose harmful properties had been destroyed, and that infection by fully virulent microbes was not necessary for immunization. This achievement required none of our present knowledge about antibodies or the cells that make them, though it did need great skill and care in choosing what materials were suitable to use. However, even the achievements of Ehrlich, Koch, Bordet and Pasteur were preceded by perhaps the most practically important of all immunization campaigns, carried out by persons who had no knowledge of immunology at all. This was the control of smallpox, usually attributed to Jenner's introduction of vaccination. His famous scientific memoir was published in 1798, but a more dangerous though effective procedure had been used widely since the 1740s. This involved using not cowpox but a weakened form of smallpox itself and was known as 'inoculation' (nowadays 'variolation'). Despite the small risk that variolation would itself cause a lethal infection, such was the prevalence and fear of smallpox in England that many mass inoculations, involving hundreds of thousands of persons, occurred during the 1760s. Razzell[3] has made a convincing case for supposing that inoculation more than any other factor was responsible for the rise in the population of this country which began abruptly during the second half of the 18th century. From his study of parish records it appears that in many parts deaths from smallpox of children and pregnant women almost ceased during this period, whereas deaths from other causes hardly changed. Since smallpox deaths made up about one-quarter of all deaths, their disappearance would be sufficient to account for the sharp population increase which was sustained, but not explained, by other changes which accompanied the beginning of the industrial revolution.

Of course Jennerian vaccination was safer, and soon supplanted variolation. It has been practised with little change ever since 1800 (this is an interesting comment on the genetic stability of the strains of vaccinia which are employed), and, together with quarantine regulations, has eradicated smallpox as a serious disease in those countries where it is widely used. However, in some of the developing countries smallpox eradication has only been undertaken seriously during the past ten years. The numbers of *reported* cases – and these will omit very many unreported – fluctuated between 65,000 and 133,000 each year from 1959 to 1968, and they are only now showing a steady decline. They are not large in relation to the total population, but they are not insignificant – and, after all, they represent people who should have had the best years of their lives before them.

It is interesting to consider what would happen if smallpox vaccination were stopped altogether. There is a clue from an unplanned experiment which occurred in the United States during the period 1919–1928, when vaccination was compulsory in some states and compulsory vaccination was prohibited in some others. In the former the incidence during this period was 6.1 cases per 1,000 population, whereas in the latter it was 115. Vaccination evidently matters!

I have dwelt on smallpox at some length, but of course prophylactic immunization has virtually abolished many other diseases which were until fairly recently large scale killers and maimers, especially of young people in Europe. Those who wish to read about the subject in greater detail may consult the account given in reference 5, 6, 7, and I will mention only a few facts here.

Diphtheria

The number of cases in England and Wales each year fell from an average of about 60,000 before 1940 when immunization was introduced to fewer than 1,000 ten years later and is now below 50. The proportion of cases which die, however, has remained about 10 per cent. This means that diphtheria has lost none of its virulence, and that if immunization were relaxed it would again become a killer.

Tetanus

Prophylactic immunization prevents tetanus completely. When persons in occupations where the risk of infection is considerable, such as farmworkers or soldiers on active service, have been immunized tetanus is virtually unknown. Yet in 1963 in rural areas of India one-quarter, and in some Bombay hospitals one-half of all deaths were attributable to this singularly painful and unpleasant disease. Since tetanus bacilli are widely present in the soil wherever there are domestic and wild animals, the risk of infection cannot be eliminated, but prophylaxis is cheap and effective. What is most needed is the organization to make use of it.

Yellow fever

This used to be the greatest killer of white people who visited tropical Africa and South America, though the native population were relatively immune. Zinsser[7] describes, for example, how in 1803 Haiti was liberated from French occupation less by the efforts of Toussaint Louverture and his followers than by yellow fever, which killed 22,000 out of General Leclere's 25,000 troops. But yellow fever vaccine has given complete and lasting protection to some 50 million persons who have been immunized with it since it was first developed in 1937.

Poliomyelitis

Between 1946 and 1958 this claimed between 3,000 and 8,000 victims each year in England and Wales. It was checked by the introduction of killed vaccine, and has been virtually eliminated since 1962 when the oral vaccine was introduced.

Those which I have mentioned are some of the prophylactics which confer virtually complete protection. There are others which for various reasons, while not guaranteeing complete immunity, confer sufficient protection to diminish the incidence of serious infections below one-quarter or less. They include whooping-cough, tuberculosis, typhoid, cholera and typhus. There is no point in extending these lists, since it can be generally stated that it is possible to develop effective prophylactics for any disease against which the natural infection gives long lasting immunity. There are always technical problems, but the main questions nowadays are whether it is worth while and who will pay.

What are the social implications? The most obvious is that immunization, comparably with improved sanitation and the introduction of antibiotics, has greatly diminished mortality and morbidity. Since I regard it as right that persons who are already alive should be as healthy as possible, this seems to me to pose no ethical difficulty. I do not imply that control of population growth is not the most urgent problem facing the human race – but in my view we must begin by ensuring that the only children born are those which are really wanted, not that fewer survive. Problems do arise in deciding on the introduction of a new vaccine, since it must be tried out on human beings before it can be proved safe and effective. These problems are similar to those facing pharmaceutical manufacturers, which Dr Hale will discuss later, but I will state briefly that within the ethical framework which obtains in this country the problems related to prophylactic immunization are primarily technical. Another and quite difficult question is to decide when and whether immunization can be relaxed. By the time that an infectious disease has almost disappeared from a community enthusiasm for continuing with prophylactic immunization against it will obviously decline. But if immunization were abandoned, an exceedingly susceptible population would grow up in which, if the disease were reintroduced, it could spread very rapidly and do more damage than ever it did before prophylaxis was begun. This means that prophylaxis must be continued – or be immediately available – unless either the disease has been totally eradicated (which is difficult to know for certain) or it can be readily cured by cheap and available antibiotics. I suspect that our children's children will be subjected to more rather than less artificial immunization than our own children, though the ways in which it is done will probably be more convenient than at present.

Cancer

Since I do not wish to steal Professor Watson's thunder, I will do no more than point out that whatever their causation cancer cells are commonly changed in a way which makes them recognizably foreign to the body in which they grow. They should therefore, according to the thesis expounded earlier, evoke an immune response. Some cancers certainly do so, but those which continue to grow obviously represent a failure of the body's immune response to eliminate them. Much research effort is going into trying to find ways of stimulating a more effective response, though so far without striking success in Man. Nevertheless, I am optimistic that immunotherapy will become a useful supplement to surgery, drugs and X-rays and may one day even be able to prevent some forms of cancer. Success in this field would hardly pose novel ethical or moral problems.

Organ transplantation

So far I have discussed situations in which it is required to increase immune responses at will; there are others in which the very opposite is needed, namely to suppress a particular unwanted or harmful response while leaving intact those which are valuable. The best known, though not numerically the most important, is organ transplantation. Kidney transplantation has a comparatively long and by now respectable history. More recently, attempts, sometimes successful, have been made to transplant livers, hearts and lungs. Doubtless before long such essential organs as thyroids, adrenal glands and pancreases will be added to the list, and we shall see attempts to graft normal stem cells to repair the deficiencies of persons born with congenital defects in their blood red cells or white cells. The obstacle to success is not lack of surgical skill but immunological rejection of the grafted tissues by their recipients. Almost all individuals have heritable differences in the structure of some of the molecules which make up the surface layer of their cells (the 'transplantation antigens'), and because of these the graft acts like a foreign antigen, and is liable to evoke an immune response which ultimately destroys it.

There are three main ways of overcoming or avoiding graft rejection. One is to ensure that the transplantation antigens of the donor and recipient differ as little as possible, by carrying out tissue typing beforehand. This requires elaborate organization, but is being done effectively on an international scale. A second is to use what are known as immunosuppressive drugs. They mostly act by killing lymphocytes or preventing them from dividing, which can certainly check the immune response against the graft but at the same time damps down the response against ordinary microbes, leaving the patient a ready prey to infections which would hardly affect a normal person. More selective immunosuppressive agents are being developed, however, which promise to overcome this disadvantage. A third is to induce specific tolerance of the foreign transplantation antigens, as I mentioned earlier. This has been achieved in experimental animals and will probably be possible, though it is not yet a practicable procedure, in man.

We can be reasonably confident that we know in principle how to overcome the immunological barrier to transplantation. In fact even with existing practices using a combination of methods more than 80 per cent of kidney grafts function well for at least two years, and many have survived much longer. This means that genuinely new ethical problems now face the public and the medical profession. They arise partly from the fact that tissues for transplantation need to be healthy and in good condition, as well as matched to the recipient. Even with speedy transport and good organization it is unlikely that enough suitably matched tissues will ever be available from volunteers at the time and place where they are needed. Possibly in the long run it will be feasible to establish 'banks' of frozen, typed and well preserved organs, taken from healthy persons killed by accident (the motor car will assist here!), but formidable technical problems must first be overcome. Meanwhile it is often a question of removing tissues immediately after death from persons who were never asked for their consent. I shall not pursue the implications of this, since they are the subject of much discussion and of forthcoming legislation.

The prevention of rhesus immunization

The one foreign graft which succeeds without intervention by a surgeon and flourishes without signs of immune rejection is the foetus in its mother's womb. We take this for granted, but it is very remarkable since the child inherits half of its characteristics from its father, and is inevitably different from its mother.

The explanation lies in the existence of a very subtle barrier within the womb, which screens the child's tissues from the mother's although the passage of nutriments and even of antibodies is not impeded. This barrier may, however, leak slightly during pregnancy and often breaks down during childbirth sufficiently for significant amounts of the infant's blood to enter the mother's circulation. When this happens the mother may become immunized against her child's blood cells and make antibodies against them; if these pass into the child at the time, or into a subsequent child, they can destroy its blood cells so extensively as to cause severe jaundice, and even death. For complex but well understood reasons only certain differences between the cells of the mother and child give rise to this sort of trouble. It occurs most frequently when the baby's cells possess and the mother's lack a factor called the 'rhesus antigen' (because it was first observed in rhesus monkeys).

It was recently discovered that immunization of a mother by her baby's red blood cells can generally be prevented by injecting into her shortly after birth a dose of human antibodies against the rhesus antigen, sufficient to combine with any of the baby's cells which may have entered into her during childbirth. Such treatment will shortly become routine, and rhesus disease will be virtually eliminated. This discovery would appear to be purely beneficial, but even it raises some interesting ethical problems.

The first arose from the fact that the number of rhesus-negative mothers who gave birth to rhesus-positive babies is quite large (about 80,000 annually in England and Wales) – though it must be stressed that less than 10 per cent actually became immunized. Human antibody is needed, and this must come either from women who have already been immunized as a result of childbirth or from volunteers who are immunized on purpose. The supply of antibody is limited, and therefore it was necessary to discover how little is needed to be effective. A careful trial was conducted in which mothers at risk received at random different amounts of antibody, including some so small that it was uncertain beforehand whether they would be ineffective or even (as was theoretically possible) actually increase the chance of becoming immunized. In the trial neither the doctor nor the women knew beforehand what dose any individual would receive. Of course, the results of treatment with the lowest dose were analysed as they came in as quickly as possible, so that this dose could be withdrawn from the trial if it was obviously harmful – though, in fact, it proved to be almost as beneficial as the largest dose used. As a consequence of this trial we now know that sufficient supplies of antibody can be made available to treat all women in this country who might need it. I think that it was wholly justifiable, but it illustrates rather well some of the problems which can arise.

The second is that if the treatment is successful it will not be many years before there will be no women left naturally immunized against the rhesus factor from whose blood antibodies can be prepared for treating other mothers. Yet the risk will not be diminished at all, even though rhesus disease itself has disappeared. This means that in the not too distant future purposely immunized volunteers *must* be used – and in fact in some countries they are used already. Not many will be needed, but such immunization will represent a definite though very small risk, even if the volunteers are males. How they should be chosen and how rewarded raises an interesting question which I will leave you to consider.

There are other and commoner conditions on which immune responses appear to be far from beneficial and sometimes positively harmful. On the one hand there are the allergies – asthma, hay fever, eczema, etc. – which arise from exaggerated responses made against what would normally be harmless materials in the environment, such as pollens and foodstuffs. They are one of the penalties for possessing an immunological system adapted to recognize foreignness in any guise. On the other hand there are the so-called 'auto-immune' diseases, in which for various reasons the body makes antibodies able to react with its own constituents, and which often involve a genetic predisposition towards failure of the mechanisms by which self-tolerance is normally maintained. I cannot dwell on these, except to state that their cure or alleviation poses similar problems in principle to those which arise from organ transplantations, and that progress in one will lead to progress in the other. Because the field of transplantation is the more dramatic, research in this field tends to attract funds more readily than in the others. If there is a lesson to be learned from this it is that the funding of research is not necessarily related to the obviousness of human needs – which should surprise no one. In relation to immunology, however, this remark applies less to government than to private funding. Not only is there a unity in this sort of research, so that advances on one front help to push forward on others, but also the fact that pharmaceutical firms are quite properly more interested in markets than in glamour provides a built-in corrective tendency towards applying useful findings where they are quantitatively most needed.

If I have tried to convince you that immunology is a fascinating aspect of biomedical science, and that its application is essentially beneficent, this is not so much because I wish to advocate the virtues of the subject on which I work as to counteract the scepticism of those who now question the value of scientific enquiry. They have a point, but they should not press it too far.

References

1 A good account will be found in 'The structure of antibodies', *Scientific American 217*, no. 4, 81 (1967).
2 Gally, J. A. and Edelman, G. M. (1970). *Nature 227*, 341.
3 Razzell, P. E. (1965). 'Population change in eighteenth-century England. A reinterpretation', *Economic History Review 18*, 312–32.
4 Parish, H. J. (1965). *A History of Immunization.* Edinburgh, Livingstone.
5 —— (1968). *Victory with Vaccines. The Story of Immunization.* Edinburgh, Livingstone.
6 Perkins, F. T. (1965). 'The science and practice of immunology', *Journal of the Royal Society of Arts 113*, 82–100.
7 Zinsser, H. *Rats, Lice and History.* New York, Bantam Books.

Discussion

Edwards I've heard people say too that they got no help from professional immunologists in the early days of rhesus immunization.

Humphrey In the case of the prevention of rhesus disease there was animal evidence that antibody against the red cells of the foetus would prevent immunization. But it didn't happen to be the rhesus group. In retrospect you can see exactly why it works and why it should go on working and you can predict how much antibody to use and so on. But your point is correct.

Hayes With respect to diphtheria, we are in a position where everybody is immune and the disease virtually doesn't exist, but in order to preserve this state of affairs we would have to maintain the immunization procedure, otherwise diphtheria might come back in a rampant form. Yet you are careful to state that you would not compel anyone to be immunized.

Humphrey You've made a perfectly good point. Although in some countries immunization is compulsory, I think there is a lot to be said for not making it so because of the slight but definite risk. The point is that herd immunity does not depend upon 100 per cent of people being immunized; it depends upon 80 per cent of people being immunized.

Monod Testing vaccines, as other drugs, is becoming a social and even a political problem, because advanced countries, beginning with the United States, are putting more and more restrictions on the official way that tests have to be performed in their own country. This goes to such extremes that some big drug companies have given up trying to do the tests in the United States and are buying the services of foreign governments to test it on foreign populations, mostly in undeveloped countries.

Humphrey In this country we have rather well defined conditions so that after experiments on animals, studies are first of all done on volunteers, e.g. in the factory which makes the vaccine. Second, the vaccine is tested on small groups which are then enlarged. But there still are problems and perhaps I could illustrate one. Rubella, German measles, is a generally mild disease, except when it is transmitted from a mother to her foetus and can cause congenital malformation. Now there is a vaccine against rubella, being developed by attenuation, which gives quite good immunity. In some parts of America it has been very widely used against a complete age group from children to young adolescents, possibly without adequate thought as to whether there will be long-lasting immunity such as is provided by the natural infection. Also we don't know whether the people who've been immunized in this way may not become subclinically infected with the real rubella during pregnancy. They could then be in a worse position than having a normal infection because they wouldn't know they had got rubella, although it might affect their baby, and they would not be alerted to go and get advice about having an abortion. In this country a relatively small group of eleven- to fourteen-year-olds, that is, we hope, younger than they can possibly become pregnant themselves, will be immunized and followed up to make sure that there isn't any development of rubella, even subclinically.

Hale I would agree with you that we have a better situation in this country with regard to introduction of new therapies than exists in the United States, and I'm speaking here as an employee of an American company. The situation here is that the Committee on Safety of Drugs act in an advisory capacity and you can reason with them. In the United States the Food and Drug Administration is mandatory, and their mandatory activities are on occasion beyond reason. What Dr Edwards said (and Dr Humphrey agreed) about the immunologist who advised against the use of Rh antibodies in rhesus treatment illustrates the danger of mandatory regulations where there is no possibility of scientific discussion and co-operative scientific decision. If that advice had been taken in regulatory fashion it might have taken some years to overcome it and some years to introduce a therapy which has been very effective. So there are dangers in regulatory mechanisms, just as there are dangers in total freedom from regulation.

Humphrey Dr Hale has made a perfectly good point. However, I could say that I don't think the immunologists ever said that the trials should not be done – they weren't asked – they were told about the success afterwards and they then began to scratch their heads to see why. Despite the natural unwillingness of scientists to give straight 'yes' or 'no' answers, their testimony tends to be dogmatic because the people who ask the questions expect a straight answer and won't accept a hedge, because they've got to take practical steps on the basis of the advice.

Eleven Molecular biological approach to the cancer problem

J. D. Watson
Professor of Biology, Harvard University;
Director, Cold Spring Harbor Laboratory for
Quantitative Biology

Every so often at a party or in a lab I hear someone say what a disaster it would be if the cancer problem were solved and there would be no money to do our science. Often the people saying this are minor jackasses, but sometimes they are really quite bright and when I hear these stale words I am more than slightly uncomfortable. But the sad fact that I know I shall hear this wisecrack still again is quite instructive about the nature of cancer research as practised today throughout the world.

Serious thinking about cancer must always start with the uncomfortable fact that most people who have been very sincerely working on cancer think that in the short term the problem is insoluble – much, much too difficult for any mortal. None the less, the financial support of much modern biology rests upon the strong desire of society to solve quickly the cancer problem. It reflects this wish by voting or giving money to support a variety of forms of cancer research. But, those people who administer this cancer money realize that there still are horrendous gaps in our knowledge of fundamental biology which have to be sorted out before you can rationally tackle the medical problem. Thus they pour much of the money earmarked for cancer research into the support of pure biology. As a result most people who have been doing the science which is supported by cancer money have no direct interest in the problem of cancer itself at all. Instead their primary aims are to solve fundamental biology problems. This is the way it should be, for if their involvement with their immediate research was only indirect they probably would not succeed, and in the long run cancer research would be greatly set back.

Fortunately the wisdom of massive support of pure biology has become easier to justify over the past few years. There is now a growing concurrence of the research problems which seem important to the scientists in university biology departments with those which seem the most crucial to the scientists who work in cancer institutes. At long last we may appreciate the magnitude of the task facing us – first in trying to pinpoint the fundamental chemical differences between normal and cancer cells, and when that is accomplished to use this knowledge selectively to prevent the multiplication of cancer cells. No one, however, should have any illusions that because of what has been accomplished so far in the development of modern biology, we can expect any rapid clinical answer. One can't just say that this problem is more relevant than many others and therefore it should be immediately solved. Unfortunately many people seem to think that this is the case, and I am sure some people suspect that most biologists are very cold-blooded for not seriously devoting their efforts to the cancer problem and instead are worrying about esoteric ideas which have no connection with real human problems. But I don't think there has been any true callousness. Many of the people who have worked on cancer because they wanted to do something for society have ended up very disappointed people, knowing that they were not up to the task. Nor for that matter has anyone so far been equal to the matter.

None the less one almost can't pick up a newspaper without reading something which seems to offer some possibility of a cancer cure. With any event which has been as cruel to a family or an individual as a case of cancer one wants continual hope. Astute newspaper editors know this and regularly print stories about recent scientific research which might somehow promote the development of a cure.

The general public thus is constantly being encouraged to believe that the next several years may at last witness the arrival of the long-desired magic weapon. They are prepared, therefore, not only to continue, but even to spend more money on cancer research. In the US there is a very good chance that the level of federal support will double if not triple over the next decade, even when many other branches of science are being cut back to the point that their survival as first-rate efforts is coming into question. Thus it is my guess that for many years there will be an excess of money for first-rate cancer research. Even projects judged marginal on scientific grounds are likely to receive support, if only because a cancer research centre within a locality gives residents the feeling that their medical treatment reflects the latest advances in fundamental cancer research

Looking back early in this century, the first real scientific discovery which had a definite therapeutic consequence was the discovery of radium and the subsequent observation that ionizing radiations of various forms kill cancer cells at a faster rate than most normal cells were killed. It was thus initially hoped that by radiation treatment we could cure cancer. This has been only partially true. It will work with some cancers but not with others. Not all types of cancer cells are more susceptible to radiation damage than normal cells and, even more important, radiation treatment generally will not work when the initial cancer has spread throughout the organism. When radiation therapy initially began, no one knew anything at the molecular level about how radiation affected cells, and even today, despite a greatly accelerated programme of biological-directed radiation research prompted by the arrival of atomic weapons, we are still much in the dark.

It was in the 1920s the great German biochemist Otto Warburg made the observation that tumour cells were characterized by the fact that they produced lactic acid in very great amounts, and that the origin of this metabolite was excessive fermentation of glucose. This discovery led to the belief that possibly there was something biochemically uniform about all cancer cells, and so vast numbers of people began studying the way glucose is metabolized both by tumours and by normal cells. This work still goes on; unfortunately it has led to innumerable controversies, some people saying that Warburg is right, others that Warburg is wrong. Many cancer institutes were filled with people doing this sort of experiment and basically to no avail. And all one can say today is that for some reason still very unclear, the control mechanisms which regulate how glucose is taken up by a cell are affected by the cancerous transformation. But whether this is a primary cause of cancer or a secondary consequence, remains unknown. None the less the lactic acid controversy provided a great stimulus towards the advancement of biochemistry. At the time Warburg made his observation almost nothing was known about what now is called intermediary metabolism – that is, the way our food is taken in, converted to smaller molecules, and then built up into bigger ones. It was a desire to solve the cancer problem which lay behind much of the growing financial support that biochemistry began to receive in the 1930s and even more so in the postwar years.

The next major discovery that was hoped to be directly relevant

Molecular biological approach to the cancer problem 143

for a cancer cure was the pinpointing of certain chemicals as agents which cause cancer. Such molecules were called carcinogens, and people in certain occupations were found to be dangerously exposed to them. The chimney sweeps in England produced one of the first understood cases – soot contains specific hydrocarbons which greatly increase the frequency of skin and scrotum cancers. The resulting field of environmental carcinogenesis has grown larger as more and more compounds are now realized to be carcinogenic with different substances specifically inducing the formation of specific kinds of cancer. Unfortunately, on the whole no one knows why. Where these molecules act in cells was, in the 1930s, complete speculation, and still remains almost completely so. If our knowledge today is a little firmer it is because modern biology has told us much about the role of DNA in chromosomes, allowing some people to think that some carcinogens act directly at the DNA level.

The main impact of the chemical carcinogen work so far has been to warn people against excessive exposure to potent carcinogens. We suspect that cigarette smoking is today's greatest danger, although there are other chemicals whose dangers still are vastly under-appreciated. One is asbestos, a very strong carcinogen, a fact many regulatory agencies would like to forget because it is a most integral component of the construction industry. Only last year did the city of New York begin preventing construction firms from spraying it into the air. But I'm afraid most other cities continue to allow this insane practice. New York City's uncharacteristically sensible action owes much to a series of articles in the *New Yorker* magazine which documented the asbestos problem in a most alarming way. But for the fundamental object of finding out what biochemically a cancer is and how to cure it, the work in chemical carcinogens has really led nowhere.

At first sight, a much more promising lead for therapeutic attacks on cancer was the observation that specific hormones frequently are necessary, not only for the growth of normal cells but also of specific tumour cells. Serious work along these lines began in the 1940s, the period immediately following the elucidation of the chemical structures of many steroid hormones. As their chemical structures became known, and they became available in substantial quantities, hopes arose that by upsetting the normal hormonal balance of a cancer patient, tumour growth could be retarded if not completely stopped. Spectacular success was in fact achieved by Charles Huggins of the University of Chicago who used female sex hormones to stop the growth of some prostate cancers. But, unfortunately, the majority of cancers do not respond well to hormone therapy, and given our still very limited knowledge at the molecular level about how hormones act, there seems little reason for more extensive success in this direction in the foreseeable future.

The next important stage in cancer research started in the late 1940s. Then the first drugs were found which killed cancer cells more frequently than they killed most normal cells. They were analogues of the vitamin folic acid, the most successful of which is called aminopterin. It was first exploited at the Children's Hospital in Boston by Sidney Farber, who observed very striking long-term remissions of certain childhood leukaemias. Unfortunately with time aminopterin ceases to be effective as resistant cells appear and almost never do complete cures occur.

Subsequent research showed that aminopterin acts by blocking the synthesis of the DNA precursor thymidylic acid, and over the next decade many compounds which interfere with normal DNA synthesis were found to cause temporary remissions of a variety of different forms of cancer. The general philosophy began to dominate cancer therapy of finding drugs which would selectively inhibit the synthesis of nucleic acids in tumours, while not inhibiting nucleic acid synthesis in normal cells. This has proved a very tall order, and on the whole it has not worked. All cells have to make DNA and RNA, and if you block the cancer cell from multiplying you generally block the normal cell from multiplying. And so all such drugs which are used are themselves very toxic, and if too much is given the side-effects can be worse than the disease.

The hope that some day a perfect drug will be found continues to generate extensive financial support for massive screening programmes which yearly test thousands of compounds for anticancer activity. Behind these searches are a number of factors. Perhaps foremost is that this way of attack has worked with the many bacterial diseases now routinely controlled by a variety of antibiotics. The thought that this approach has once worked so well against bacteria carries more weight than the equally striking fact that no really effective antibiotic against viral agents has yet been found – despite very very intensive screening throughout the pharmaceutical world. There is also the fact that two rare forms of cancer, choriocarcinoma and the Burkitt lymphoma, can often be cured by specific drugs. So there may not exist any intrinsic reason why some day a perfect drug, say one applicable to breast cancer, might not show up. There is also the consideration that since many forms of chemotherapy are now routinely used in hospitals throughout the world, there is a tendency to believe that they are more effective than they really are – a viewpoint which leads to the expectation that only slight improvements need to be perfected before drugs of the type we now use would really be effective. And finally, the current screening programme, largely done in the United States, yearly costs only some $30,000,000. In one way this sum is quite large, say in comparison to that spent on pure research on animal viruses. Yet it is insignificant with respect to that used to produce weapons systems that never work.

So even though most people who have long been involved with current anti-cancer screening programmes are very discouraged and many of the brightest people want to stop them, I suspect we shall continue to screen a large variety of the new chemicals which become available each year. Of course this may be regarded almost as a superstitious response. But since our fundamental scientific knowledge is so inadequate, the prediction of 'experts' is hard for many people to accept, especially if you know someone who is dying and desperately want something to turn up.

A very different approach comes from the study of the viruses which can cause cancer in a great variety of animals. Though the first cancer virus, the Rous sarcoma virus of chickens, was discovered in 1912, only during the past decade has this field been a main focus of cancer research. There were many reasons for this neglect. For one, there is abundant evidence that cancer is not normally an infectious disease. In the second place, when cancer cells are examined with the electron microscope, only rarely are particles seen which might be virus particles. And perhaps most importantly, no human cancer has yet been shown definitely to be caused by a specific virus. But now the atmosphere is changing very rapidly, largely because of growing evidence that viral-induced cancers in animals like the mouse are not the exception, but in fact may be the rule. For example, as leukaemia in more and more animals is being shown as a viral disease, it becomes harder and harder to avoid the idea that human leukaemia must also be viral-induced.

The knowledge, however, that most cancer virology now centres on tumours produced in laboratory animals bothers those people who would only be excited by this field if the viruses studied were known to cause human cancers. But to me the sanest approach now available is to concentrate on the *simplest systems*, where the change from normal to malignant can be biochemically analysed. Hopefully, when such systems yield fundamental insights, the problem of human cancer will be seen amenable to rational attack. If on the contrary we were *now* to spend most of our efforts on brute-force attacks to show that many human cancers are viral-induced, we are likely to end up only with massive confusion.

Biochemical work with viral-induced animal cancers now rests on a firm experimental foundation because of two very important technical advances of the past decade. One is the ease by which a variety of animal cells can be grown outside living animals – a technique that is called tissue culture. Cells in tissue culture can grow as single cells as though they were bacteria, thereby offering the possibility of well controlled studies of their biochemistry. The second main advance is the finding that viral-induced carcinogenesis can *routinely* be induced in such tissue-culture-grown cells. As few as twenty-four hours after the addition of a Rous sarcoma virus to a tissue-culture-grown chicken cell, the external morphology of the cell changes from normal to malignant. In contrast, only some fifteen years ago all biochemical studies had to be done on tumours growing in animals. Under these conditions it was impossible to decide whether a biochemical change specific to a given cancer cell was intrinsic for the cancerous transformation or whether it was an irrelevant change which arose much later when the tumour was growing in the animal.

Use of model viral systems offers another real advantage: if you look at the complexity of the human cell and ask how many different genes are present in a given cell, one comes out with a horrible answer. The number of genes may be greater than 100,000 and might even be a million. Of these we have knowledge today of only very very few. If a mutational change in only one of these genes was responsible for a specific cancer, it would be like looking for a needle in a haystack – where would you know where to look? You might go on working at our current pace for a hundred years without getting anywhere near the answer, even if you worked most intelligently and sensibly.

Recent work on viral carcinogenesis has made this frightening possibility much less likely. Over the past few years it has become very clear that viruses cause cancer in a very *direct* way. The chromosome of the virus enters the cell and, through a process of genetic recombination, inserts itself into a chromosome of the host cell. There its genes continue to function, with *one* of them being the direct cause of the cancerous change. Before this point had been established, there were fears that perhaps cancer viruses acted in a 'hit and run' way. Under this hypothesis the virus got in the cell, randomly damaged it somehow to cause the cancerous transformation, and then disappeared. If this latter hypothesis – that cancer viruses, like radiation, caused damage randomly – had been correct, we would be in a bad way. Viruses would just be another carcinogenic agent which we didn't know how to attack in a biochemical fashion.

Even more interestingly, you can induce cancers by viruses which are among the smallest known. The best understood group of cancer viruses, one of which is a mouse virus called polyoma, another, a monkey virus called SV40, only contain somewhere between five and ten genes. One of these genes, when it gets on to a chromosome, causes the cancer. So conceivably instead of having to work out the functional task of all the 100,000 or more human genes in order to find the one which is responsible for the cancer, we may be able to limit our attention to the much much smaller group of viral genes. By characterizing the viral multiplication cycle very completely, and in doing so, identifying the exact functions of each viral gene, we should be able to pinpoint the exact biochemical process at the root of the change to a cancer cell. Thus the use of viral systems has changed our feelings about the difficulty of the cancer problem by many orders of magnitude.

The complication that tumour viruses can contain either DNA or RNA as their genetic material is not as serious as first thought. With the viruses that contain DNA (the polyoma virus is one such) it is very easy to imagine how genetic recombination inserts a viral chromosome on to a chromosome of the host cell. But for many years, the dilemma existed of how to deal with many tumour viruses which used RNA as their genetic material. Among these RNA tumour viruses, all of which superficially look like influenza virus, are those which cause leukaemia in a large variety of animals, as well as breast cancer in mice. If their genetic component was always RNA, there appeared to be no easy way for them to become part of a host chromosome, the essential component of which is always DNA. Then, some five years ago, Howard Temin of the University of Wisconsin proposed that the first step in the RNA-virus-induced conversion of a normal cell to a cancer cell was the use of the viral RNA to serve as a template to make a DNA strand of complementary nucleotide sequence. This DNA would then integrate itself into a host chromosome.

Though Temin's ideas were seriously considered, they were at first thought unrealistic. No one seriously liked the idea of the DNA→RNA story being reversed, especially since Temin's main experimental evidence was always borderline. Then about six months ago he, and independently David Baltimore working at MIT, discovered that within the tumour virus particles themselves is an enzyme by which the RNA is transcribed into DNA. The existence of this enzyme, now sometimes called 'reverse transcriptase', can only mean that Temin's ideas are essentially correct. Thus, even though all the enzymatic experiments are not yet completed, we now think we have the correct theoretical framework for determining how these RNA tumour viruses can get on to host chromosomes.

Over the next few years, our main attention is likely to turn to the study of the functioning of these viral genes when they reside on host chromosomes. Most important, the two main characteristics of cancer cells now cannot easily be related to each other. One is that they divide when they shouldn't. In a superficial sense you can say they make DNA when other cells do not make DNA. So there seems to be some failure of the normal control of DNA synthesis. The second main characteristic of tumour cells is that their surfaces are so modified that they lose their ability to stick to the correct cells. Somehow their surfaces don't have the right affinities, and they move to regions where their normal counterparts cannot survive. This poses the dilemma of how a change in a single viral gene can lead to two such apparently unrelated consequences: a failure of DNA synthesis to be controlled and a change in the surface which produces altered cellular affinities.

A very fundamental obstacle to successfully attacking this problem is the fact that the molecular structures of the surface membranes of normal cells are completely undeciphered. In the sense that it would not have been possible to do molecular genetics without knowing the structure of DNA, it may not be possible rigorously to know the essential differences between normal and malignant cells until an immense amount of chemical effort is done to elucidate the exact structure of the various

Molecular biological approach to the cancer problem

chemical groups on the surface of cells. When you ask why people haven't yet accomplished this objective, the answer is straightforward. The problem has been far too difficult for any single individual or group of scientists working in a single laboratory. But as almost every second biochemist now speaks about concentrating his research on cell membranes, hopefully enough people in the next decade will have the guts to follow up their words. But because it's so tricky, most who will try, will fail. And if the chances of failing are too great, intelligent scientists will stay away and the only people who will be at it are those who don't know what's up. So we just have to hope that someone really clever will join in.

Fortunately there exist some new empirical observations on the cell surface which already may have very great consequences. They arose from a chance observation by the pathologist Joe Aub who, working at the Massachusetts General Hospital, searched for an enzyme which would selectively kill cancer cells. In doing so, he added an extract of plant tissue (wheat germ) which contained an enzyme called lipase, and found that it selectively agglutinated cancer cells. Later, it turned out that it was not the enzyme that agglutinated the cancer cells, but an impurity, shown by Max Burger at Princeton to be a plant glycoprotein, that has selective affinity with the surface of cancer cells. This wheat glycoprotein, and also the jackbean protein conconavalin A, each has two identical combining groups capable of specifically binding to the surface of a cancer cell. Therefore, when added to cancer cells, each molecule can link together two cancer cells and agglutination results. But gently treating conconavalin A with the enzyme trypsin makes it able to stick to only one cancer cell. Even more important, when such modified conconavalin A is added back to cancer cells growing in tissue culture, the cancer cells begin to act like normal cells.

To explain his findings, Burger postulated that change from a normal cell to a cancer cell involves exposing of chemical groups which are normally masked. That is, part of the outer covering of normal cells is effectively removed when the cancerous transformation occurs. This removal exposes a chemical group on the inner surface which specifically combines with the plant glycoprotein; by adding an excess of the glycoprotein, the inner surface is again covered up, thereby making it superficially resemble a normal cell, particularly in the re-establishment of the normal control mechanisms regulating DNA synthesis.

These most exciting findings immediately raise the question: Can't we cure cancer by adding those glycoproteins which specifically attach to the surface of the cancer cells and generate a signal which stops DNA replication? Now, unfortunately, we suspect this approach will not work in a straightforward way. After the injection of large amounts of a glycoprotein into a cancer victim, an immunological response to the glycoprotein most certainly would soon nullify its effect. But perhaps some day we may be able to find compounds which would not generate an immunological response, yet somehow would specifically cover up the crucial portion of the cancer cell surface. Thus I would think there is no question but that Burger's work will quickly be followed up throughout the world.

To conclude, I think the feeling of many people is that this next decade may possibly witness the unravelling of the fundamental chemical events which change a given type of normal cell into a cancer cell. Success clearly will demand much luck and a great deal of persistence. We can only hope that, when this understanding occurs, soon afterwards there will be a clinical consequence. But now to worry whether this will be the case probably makes no sense, if for no other reason that it will divert us from the immediate challenges that we must somehow surmount.

General references

Burnet, F. M. (1957). 'Cancer: biological approach I. Processes of control', *Brit. med. J.* i, 779. A very thoughtful article which concludes that the control of cancers that arise from somatic mutations will be very very difficult. Though now much out of date, this article still remains most worth-while reading.

Warburg, Otto (1956). 'On the origin of cancer cells', *Science 123*, 309. The lactic acid situation as seen through the eyes of its discoverer. Very one-sided and out of date, but no one should try to understand the various controversies without starting with Warburg's own expression of where the cancer field stands.

Huggins, Charles (1967). 'Endocrine-induced regression of cancers', *Science 156*, 1050. A summary of the author's imaginative work of the use of specific hormones to inhibit tumour growth.

Glemser, Bernard (1969). *Man against Cancer*. New York, Funk & Wagnalls. An excellent book for the lay reader containing a good account of where current chemotherapy screening programmes now stand.

Andrews, Christopher (1970). *Viruses and Cancer.* London, Weidenfeld & Nicolson. A popular account of current work with tumour viruses.

Watson, J. D. (1970). *Molecular Biology of the Gene,* 2nd ed. New York, W. A. Benjamin. The last chapter discusses recent experiments aimed at elucidating how tumour viruses act at the molecular level.

Burger, Max, and Noonan, K. D. (1970). 'Restoration of normal growth by covering of agglutinin sites on tumour cell surfaces', *Nature 228,* 512. The latest paper about the amazing specificity of plant glycoproteins for cancer cell surfaces.

Discussion

Adinolfi Do you think that studies on the mechanisms which induce synthesis of foetal antigens to be switched on* may help in explaining the origin of cancer cells and their abnormal proliferation?

Watson The question relates to the fact that if you look by immunological techniques at cancers, you often detect a resemblance to proteins which are present in early embryological development. Now recently there has been a claim (it is still too early to know how solid it is) that the immunological surfaces which one is looking at are actually coded by one of the leukaemia viruses that one can detect in solid tumours. Thus, viruses which cause cancer in later life may have some other type of function in embryological development, and this may be the origin of these antigens.

Dewey I was very interested in this very lucid exposition of the cancer problem. But as a cancer worker myself, can I take it one stage further and say the real problem is that cancer cells have lost their control mechanism. Maybe the virus is adding not a gene but a suppressor substance. It seems to me that in the cancer problem, we should now be looking at these control mechanisms for cell division or cell differentiation, and this is a very difficult problem. It is possible that many cells in contact will inject these control substances into each other and, therefore, a change in cell surface might well stop this

* See *Nature* (1970). *228,* 608.

injection. You can show this mechanically by putting a little bit of plastic in tissue; you then get cancer cells growing up against the plastic. These cells have been mechanically removed from the control substances. So, while I think there is a tremendous need to look at these cell surfaces, it may well be a totally irrelevant part of the actual cancer problem itself, which is lack of the control mechanism.

Watson I think you are wrong. If one takes Burger's observations as solid, the primary event is a change in the cell surface which leads to some loss of control of DNA synthesis. That is, the surface is changed first. The other possibility was that you first changed the control mechanism which then led to changes at the surface. Up to now, there has been virtually no reason for choosing one idea over the other. The work of the past year makes me strongly believe that you should concentrate first on the surface.

Channon I wonder if Professor Watson would make any comment about his remarks concerning the use of money, because it seems to me that he has washed his hands of educating the people who are voting the money. Is there any hope for science research to be directed by scientists in the future?

Watson American society as a whole is going to decide how to spend its money as long as the United States remains a democracy. Considering the reasons that generate cancer research, it's better to be positive and say there is something we want to have done and try to get money for it, rather than to stop those types of experiments which our intuition tells us will get nowhere. Of course, if I knew of good clinical research in the cancer area which would not be supported because of absence of money, then I would yell very loudly. Unfortunately, as I implied above, good ideas are very scarce indeed. So, if they soon vote a billion dollars in the American Congress for cancer research, and if only 10 per cent of it could be spent wisely, then this total would still be more wisely spent than most money spent in the United States, and I think we should be thankful.

Sussman I object to the 'curing of cancer' as the scientists' war-cry for money, on several grounds. First, I object to the hypocritical and cynical attitude of scientists who want to help the public but who betray their contempt for this same public when they say the public must be fooled into giving them money to do good research. Second, emphasizing the utilitarian role of science causes scientists to behave falsely. Their proposals are doctored to make them 'relevant', and political games develop among scientists for the sharing of the available money. Thus, this attitude is both corrupt and corrupting. Third, on purely practical grounds, basing support of science on the accomplishing of specific goals may prove self-defeating as shown by the moon-programme.

I believe support of science can most honestly be based on the proposition that science, like art and the humanities, is a primary aspect of our culture which can, on occasion, produce tangible benefits. This may not be very practical, but I believe the utilitarian approach may be even more impractical.

Part five Agricultural botany and the environment

Twelve Molecular biology and agricultural botany

Arthur W. Galston
Professor of Biology
Yale University

The science of botany has remained relatively untouched by the remarkable recent advances in molecular biology. This is due, at least in part, to the traditional conservatism of botanists, but also in some measure to the relative intractability of plant cells to many of the kinds of experiments performed in molecular biology. The plant cell lives inside a wooden box, the cell wall, which contains cellulose pectins, lignins and other yet undefined polymers. This rigid wall greatly complicates the extraction of materials from cells, the introduction of materials into cells and experiments involving cellular fusion. The cytoplasm is restricted to a narrow layer pressed between this rigid wall and the large central vacuole, making localization and autoradiographic techniques difficult. The vacuole, which occupies about 90 per cent or more of the volume of the cell, frequently contains noxious materials such as tannins, acids and phenols which denature proteins and damage cellular particulates that one wishes to extract. The plant body is also relatively undifferentiated biochemically, with numerous functions residing in each cell. This greatly complicates the isolation of separate biochemical pathways and systems. Despite these difficulties, some remarkable advances have recently been made, and certain of them may well produce dramatic innovations in agriculture botany in the near future.[1]

With the world's population currently at 3.5×10^9, and increasing at an annual rate of about 1·8 per cent, the equivalent of sixty million new mouths to feed have been added this year. The best lands of the world are already under cultivation, with less than one acre of crop land and one acre of pasture or meadowland for each person on earth. Improvements in our understanding of factors controlling crop production are absolutely essential if we are to avoid a catastrophe stemming from widespread starvation. It goes without saying that no increase in food productivity alone can solve this problem, which must ultimately be linked to stabilization of the world's population through the co-operative action of all governments.

Modern intensive agriculture brings with it many technological problems. The highest yielding genotypes require massive applications of fertilizers, especially those containing nitrogen and phosphorus. Inefficient utilization of these plant nutrients leads to the runoff of significant quantities of nitrates and phosphates into water systems. Eutrophication of lakes and build-up of toxic quantities of such derivative products as nitrite may result. If it were possible to improve biological nitrogen fixation, for example, a part of this problem might be obviated.

Succulent, high yielding strains of crop plants offer tempting targets for predatory insects, fungi and bacteria, and competition by weeds. Man fights these pests by massive applications of toxic compounds in the form of insecticides, fungicides and herbicides. In the US alone, we spend about $2 thousand million annually for such chemical pesticides, many of which, like DDT, are persistent in the environment and deleterious to certain aspects of the ecosystem. Are we for ever doomed to this practice of polluting our environment with noxious organic pesticides in order to insure an adequate crop yield? It is my belief that we have paid insufficient attention to the natural defence mechanisms of plants against their predators, and that scientific research can indicate new pathways to disease resistance which are not dependent on the application of chemicals. I hope to describe some of these possibilities briefly in the course of my paper.

Vegetative propagation by single cell culture

One area in which botanical research has proceeded rapidly is in the culture of single cells, and in an understanding of the exact nutritional requirements for their potentially limitless growth and their controlled differentiation in chemically defined media. It has long been known that the higher green plant is a complete autotroph and, when supplied with carbon dioxide, radiant energy, water and minerals, can complete its entire life cycle. No exogenous vitamins, amino acids, hormones or in fact any organic molecules are required. Various parts of the plants, however, are heterotrophic for one or more growth factors.

Thus, excised roots may be grown in pure culture only if they are supplied with thiamin; some also require nicotinic acid and pyridoxin. Cells of stem tissue do not require vitamins, but require external supplies of two hormones, auxin and cytokinin produced, respectively, in stem and root apices. Auxins are generally derivatives of 3-indolylacetic acid (IAA) or the free IAA itself; cytokinins, on the other hand are mainly N-6 isopentenylated derivatives of adenine or related purines.

The relative concentrations of auxin and cytokinin determine the morphogenetic fate of cells in culture. Thus, roughly equimolar auxin and cytokinin leads to rapid cell division and the production of undifferentiated callus, while a molar excess of cytokinin leads to the initiation of formed buds and an excess of auxin leads to the initiation of formed roots. Thus simple control of the hormonal components of the nutrient medium permits the experimenter to reconstitute an entire normal plant from a single, normally non-dividing cell removed from almost anywhere in the plant. This experiment is one of the most elegant demonstrations of the fact that the totipotency of the genome remains intact in the differentiated cells of the organism.

Our ability to propagate entire plants through culture of explanted vegetative cells gives us a powerful tool for the maintenance of desired genotypes which might otherwise be lost because of failure of the normal sexual reproduction mechanism leading to seed formation.

Cytokinins, protein synthesis and senescence

Cytokinins have some relevance for senescence phenomena in plants. The yellowing of senescent leaves on a plant can be delayed by the application of cytokinins, whose effect seems to be related to enhancement of the rates of RNA and protein synthesis and a channelling of mobile nutrients to the site of hormone application. These effects also explain the previously mysterious decline in rates of RNA and protein synthesis of leaves detached from the plant. Since these phenomena are similarly slowed by cytokinin application, senescence in leaves may reasonably be interpreted as being due, in the first instance, to a decline in cytokinin titre.

Cytokinins are not limited to higher plants; several are found in yeasts and bacteria, especially in the transfer-RNAs for a group of amino acids, including serine, phenylalanine, tryptophan, tyrosine and leucine, which have U as the first letter of the triplet codon. Sequential analysis of such t-RNAs reveals that the cytokinin is adjacent to the anticodon triplet and it has been reasoned that this must have some significance for the alignment of transfer-RNA and messenger-RNA. If this is true it is probably the most elegant delineation yet of the exact role of a hormone in the control of the synthesis of proteins. Cytokinins have also been shown to attach to higher plant ribosomes.[2]

I should add that there are arguments against the interpretation that cytokinins function by virtue of their incorporation into t-RNA. For one thing, there seems to be no codon for the N-6 isopentenylated adenines, and, in fact, experiments with double-labelled molecules indicate that the cytokinin nucleus and the side-chain are introduced separately into the t-RNA. Labelled mevalonate and other isoprenoid precursors yield the isopentenyl side-chains for attachment to the 6-amino group of adenine already incorporated into the t-RNA, a process analogous to the well-known methylation of bases in formed RNA. It also appears that cytokinins may be active without becoming incorporated into any high molecular weight materials, since radioautographic evidence indicates that the label can be washed away some time after application without any loss of the morphogenetic action. Whatever the mode of action of the cytokinins, a knowledge of their chemistry has given powerful tools to the botanist for the study of protein and RNA synthesis, senescence, and the control of growth and morphogenesis. The practical consequences of increasing the longevity of harvested parts of plants is obvious. Spinach leaves, broccoli heads and carrot roots, for example, might be preserved for considerably longer times after harvest following cytokinin application.

Haploid plants

The plant geneticist normally begins his operation by inbreeding a desired progenitor for a sufficient number of generations to ensure relative homozygosity. This is a slow and tedious process which could be circumvented by the artificially stimulated development of a haploid germ cell into a plant, and a subsequent doubling of the chromosome number by colchicine or similar mitotic agent. This possibility was foreseen in 1941 by van Overbeek and collaborators who tried to aspirate and cultivate on artificial media the unfertilized eggs of *Datura*. But this was in the pre-cytokinin days, and the nutrient media employed were insufficient to the task. Even coconut milk, the nutritional medium *par excellence* of plants, which is now known to contain cytokinins and other growth factors, does not do the job. This problem is still an enterprise worthy of attention.

Recently success in the cultivation of haploid plants has come from another source. Nitsch and his colleagues in France have shown that by immersing the young anther in a nutrient medium containing abundant supplies of the natural growth regulatory hormones and other substances, anthers containing immature pollen grains may be induced to form haploid plants. The resulting cell divisions produce haploid callus tissue which can be induced to form a normal plant. This practice gives promise of greatly accelerating plant-breeding programmes. Its accomplishment would have been impossible without precise knowledge of the molecular requirements for growth and differentiation.

Protoplasts and somatic genetics

It has now been convincingly demonstrated that animal cells of widely divergent genomes can be caused to fuse somatically, and that genetic material from both parent cells may persist, at least for some time, in the fusion product. Such work with plant cells has been hampered by the presence of the cell wall. Now, however, techniques are available for the removal of cell walls and subsequent fusion and regeneration of new cell walls, via isolated protoplasts and their fusion products. Multicellular plant masses may first be separated into single cells by the use of pectolytic enzymes which dissolve away the intercellular cement. After this, cellulase or other beta-D-glucanases may be used to digest away the external cell wall. If the cellulase is pure, especially if uncontaminated by nucleases, if the period of exposure to the enzyme is short, and if the resulting cell-wall-free protoplast is osmotically stabilized in hypertonic sucrose, mannitol or salts, then viable products may be obtained. In several laboratories such protoplasts have been observed to stream, to respond to hormones, to regenerate cell walls, to replicate viruses and to fuse.[3] So far, fusion products have not been carried to the point of regeneration of tissue masses or of new organisms, but this field is in its infancy and one may expect rapid progress in the years ahead.

What hopes does this new somatic genetics of higher plants

Molecular biology and agricultural botany 159

hold for us? One can dream of many exciting possibilities. Consider, for example, a successful somatic fusion between a leaf cell of a cereal, such as rice, wheat or maize, and a cell from the nodule of a leguminous root which has been infected with nitrogen-fixing bacteria. If one could regenerate such a cereal plant with nitrogen-fixation ability, its yields could be improved all over the world without the onerous and expensive necessity of heavy nitrogen fertilization. Consider also the impact of the introduction of genes for resistance against fungi or other pathogens which cannot be normally introduced by genetic means because of pollen incompatibility or of accidental characteristics of the reproductive mechanism.

Artificial introduction of informational RNA

What we know about the synthesis of proteins in plants tells us that the basic mechanisms delineated for micro-organisms also operate in these eucaryotic cells. It therefore stands to reason that the introduction of informational RNA from outside a cell could be used as a tool to control the synthesis of desired proteins not normally coded for by the genome of the cell being employed. We have reason to believe that we have been able to accomplish this in our laboratories at Yale.

Tobacco pith cells *in situ* contain a low level of peroxidases. When a block of pith tissue is explanted aseptically to a suitable environment, it rapidly forms new enzymes, including peroxidase. Electrophoresis on starch gel reveals that while the parental peroxidase exists entirely as two anodic isozymes the newly formed peroxidases are all cathodic. It appears that the ability to form cathodic isozymes is repressed *in situ* and for some reason derepressed after some time *in vitro*.

In examining the possible basis for such control we reasoned that a repressor protein formed *in situ* might be responsible for the lack of formation of the cathodic isoperoxidases. In support of this notion we have found that RNA extracted from fresh pith tissue and infiltrated into such tissue cultivated *in vitro* prevents the formation of the new isozymes. Such activity is destroyed by hydrolysis with KOH, and partially by treatment with crystalline ribonuclease. RNA extracted from pith first cultured for twenty-four hours *in vitro* shows much less repressor activity, and pith cultured for forty-eight hours *in vitro* is totally without such repressor activity. Thus, it appears that an RNA present in fresh pith prevents the formation of the cathodic isoperoxidases, and that the high titre of this compound in the tissue falls when the tissue is removed from the plant.

Fractionation of the effective RNA reveals that ribosomes are totally without activity, and that when the soluble material is separated on MAK columns and analysed in typical pulse-labelling experiments, it has many of the characteristics of messenger-RNA. We thus come to the tentative conclusion that we have been able to extract and introduce an extracted messenger into functioning plant cells.

The possible implications of such a finding are immense. With respect to peroxidase, for example, there is in many plant tissues an inverse relationship between growth rate and peroxidase activity. If this correlation is meaningful it should be possible to control the growth of cells by the introduction of informational RNA which leads either to induction or repression of specific peroxidases. Since high peroxidase titres have also been correlated with disease resistance in plants, such informational RNA might be used to increase disease resistance in agriculturally important plants. Other enzymes, with other important practical implications, might be controlled by similar means. If disease resistance can be introduced in this fashion, then the use of chemical pesticides might be accordingly diminished, with a resulting decrease in pollution of the ecosystem.

Possible transformation in a higher plant via extracted DNA

One obvious way to alter the genotype rapidly is to introduce effective DNA from outside the cell. Hess[4] has recently reported some experiments with *Petunia* which indicate the successful transfer, via extracts, of DNA coding for anthocyanin formation. If white-flowering mutants are treated with DNA extracted from plants of their own genotype, then about 5 per cent of the recipients show slight anthocyanin production, 95 per cent remaining white. This slight tendency to anthocyanin formation was unstable, both in sexual crosses and in vegetatively propagated offspring. When, however, the DNA was extracted from red-flowering plants, 18–25 per cent of the recipients developed anthocyanin, depending on the stage of their treatment. Such anthocyanin production appeared to be stable, both during vegetative propagation and after selfing.

The effectiveness of the extract was lowered by treatment with

DNAse, but not RNAse, indicating that the effective material was in fact, DNA from the donor plant. Gel filtration experiments indicated a minimal molecular weight of 200,000 for the effective molecule, which, on the basis of a strong hyperchromic effect, appeared to be predominantly native.

If this simple experiment can be repeated and extended, then the regular transformation of plant cells should become a possibility. Cultured single cells or protoplasts should transform more effectively than organized entire receptor plants, and if such treated cells or protoplasts can then be successfully programmed to develop into entire, mature organisms, many desirable genetic characteristics could be incorporated into their developmental patterns. Crop plants could be affected in all sorts of desirable ways by this technique, which is obviously only in its infancy.

Photorespiration and increased yields

It has recently become clear that the over-all yield of agricultural crops is the result of not only their rate of photosynthesis, but also the rate at which photosynthate is wasted during respiration. It appears that many tropical plants like sugar cane, which can fix carbon into photosynthate more efficiently than any other plants, differ from temperate plants in that they have a relatively low rate of respiration, which is not greatly affected by a rise in temperature or by strong irradiation with visible light. The extra burst of CO_2 output in the light, called photorespiration, is believed to involve glycolate as a substrate and a flavin enzyme which catalyses its oxidation. Recently developed techniques for the mass isolation of cellular particulate fractions have yielded evidence that a discrete new organelle, the glyoxysome, rich in glycolate and its flavin oxidase, may be the seat of this photorespiration. When photorespiring plants, such as tobacco, are treated with alpha-OH sulphonates, which selectively inhibit glycolate oxidation, then photorespiration is sharply decreased.[5] Clearly, this could give rise to sharply increased yields by virtue of an inhibition of photorespiratory wastage of substrate. An unexpected benefit of the application of such reagents is that they also bring about partial or complete closure of the stomata, and thus offer a means of retarding water loss through these pores. In arid zones, this could mean the difference between life and death for a growing plant. Thus, the study of photorespiration may one day yield great benefits to man in the way of increased yields of important temperate zone crop plants.

Phytochrome and the control of plant behaviour

Many plant responses, from seed germination to flower initiation, and including stem and leaf growth, are light-dependent and controlled by a pigment called phytochrome, which exists in most plant cells in minute quantities. This chromoprotein, with a molecular weight of about 120,000 and an open-chain tetrapyrrole chromophore, can exist in two mutually photoreversible forms, P_r and P_{fr}, absorbing at 600 nm (red) and 730 nm (far-red) respectively. Each can be converted to the other by absorption of a quantum in its peak absorbing region. Normally, light converts P_r to P_{fr}, and in darkness there is a spontaneous reversion from P_{fr} to the more stable P_r form. This cyclical transformation during day and night is linked to the specific photoperiodic requirements for flowering through an interaction with endogenous rhythms.

Clearly, if one could stabilize phytochrome in, let us say, the P_{fr} form, one could control numerous processes in the plant at will, and free the plant from dependence upon particular climatic conditions, such as long days for flower initiation. Recently, the P_r form has been shown to be more sensitive than the P_{fr} form to attack by aldehydes. After exposure to aldehydes, it loses its photoreversibility; the P_{fr} form is not affected in this way. The difference is apparently due to two extra lysine residues exposed during the $P_{fr} \rightarrow P_r$ photoisomerization. I predict that a greater understanding of this phenomenon will lead to our eventual ability to control at will many light-mediated plant processes which are currently beyond our capacity to affect in any way.

Newer plant hormones

In addition to cytokinins and auxins, the plant physiologist has recently discovered the existence of several other types of plant hormones which greatly extend his range of understanding and control of the plant. Gibberellins, which control seed germination, stem lengthening and floral induction in long-day plants, apparently operate via a genetic derepression mechanism, leading to enhanced enzyme synthesis. Ethylene, which controls fruit ripening and leaf fall among other processes, also induces enzyme synthesis, possibly via effects on membranes of compartmentalized systems. Finally, the isoprenoid substance abscisic acid, which probably controls the onset of dormancy in woody plants and seeds, acts to prevent derepression by such agents as gibberellin, thus effectively shutting down metabolic systems. Each of these substances has already found some commercial use, and their influence on agriculture can only grow in the future.

Thus while the food crisis facing man is serious and growing, our increasing knowledge of that wonderful machine, the green plant, may yet permit man to gain time in his struggle against starvation, pollution and waste.

References

1 A general reference for almost all the subjects discussed in this essay is Galston, A. W. and Davies, P. J. (1970). *Control Mechanisms in Plant Development*. Englewood Cliffs, New Jersey, Prentice-Hall.
2 Berridge, M. V., Ralph, R. K. and Letham, D. S. (1970). 'The binding of kinetin to plant ribosomes', *Biochem J. 119*, 75–84.
3 Cocking, E. C. (1970). 'Virus uptake, cell wall regeneration and virus multiplication in isolated plant protoplasts', *Intl Rev. Cytol. 28*, 89–124.
4 Hess, D. (1969). 'Versuche zur Transformation an höheren Pflanzen: Wiederholung der Anthocyn–Induktion bei Petunia und erste Charakterisierung des transformierenden Prinzips', *Ztschr. für Pflanzenphysiol. 61*, 286–98.
5 Zelitch, I. (1966). 'Increased rate of net photosynthetic carbon dioxide uptake caused by the inhibition of glycolate oxidase', *Pl. Physiol. 41*, 1623–31.

Discussion

Giles Although the quantity of proteins in plant seeds can be high, the quality is deficient because some of the amino acids which we require, but cannot synthesize in our bodies, are either absent or present in very low quantities. Does Professor Galston think there is any chance that new seed protein cistrons could be introduced into these plants to give them an improved amino acid composition and one perhaps even better than we at present get in animal proteins?

Galston My answer must obviously be a guess. I think that
we are barely starting in the matter of the transformation and
transduction of plant cells. The possibilities you mention
could be achieved. The new cistrons could be from the genome
of other organisms or they might be synthesized specifically
for this purpose.

Corbett You were extremely anti-pesticide, but as far as I
could see you did not offer a technical alternative which
is immediately applicable to the problems which pesticides are
used for. I wonder if you could do that and give us a realistic
time-scale.

Galston When you say that I am anti-pesticide, you mistake
my stand. I recognize our current absolute dependence on
pesticides in a sense; we are trapped into having to use large
quantities of these substances. Sometimes I think we use too
much in the way of pesticides. I know many farmers who spray
religiously on a frequent schedule suggested to them by the
chemical company which sells the pesticides, simply on the
basis that they should not wait until the first insect or fungus
is seen. I think we could more rationally use our pesticides
and could effect a great reduction in the chemical incursion
into the ecosystem simply by that revised practice. But at
present I think we are hooked; we are forced to use pesticides
because of our intensive agriculture.

With reference to my hopes for new techniques in combating
plant predators, we are starting to know a little bit about the
defence mechanisms that plants have against invading micro-
organisms. It is perfectly clear that in some instances particular
phenolic compounds generated within the plant are related
to disease resistance. If their level is kept high, genetically or
physiologically, pesticide use might be lessened. There is also
a new group of compounds called phytoalexins which seem to
be induced as a result of invasion by a particular micro-
organism. These compounds result in a measure of what
might be called immunity, although this does not involve
an immune protein mechanism in the sense that Dr Humphrey
was discussing. Also, we are starting to learn that some
enzymes, like the peroxidases, are associated with the prevention of the invasion of plant cells by mycelia of certain
phytopathogenic fungi. With this increased knowledge and with

Molecular biology and agricultural botany 165

the ability to control enzymes or biosynthetic pathways by the introduction of informational nucleic acids from the outside one will start to have an ability to get plants into a fairly resistant shape without the massive use of pesticides.

When you asked for a realistic time-scale, all that I can say is that it will depend on the level of support for research which goes in this area. If the present trend in the United States is followed, and if the scientific establishment is progressively dismembered, it may take several millennia before we can reach that happy state. But if in fact we do have adequate funds, I can foresee it within a generation.

Corbett If you imply that you've got to use some kind of genetic manipulation, your technical problem in producing plants resistant to disease or insects or whatever, is that the plants are not going to be able to respond quickly enough genetically to changes in the genome of the pest.

Gorinsky As a botanist, I should like to take up the point on specification. At the moment we are dealing with the problem of food and essentially we are dependent totally on plants for food. Also we are faced with the crisis of a loss of speciation, a loss of diversification, and we are about to see perhaps the extinction of about 800 species.* I feel that the social and ethical problem here is not in terms of the green revolution producing a 25 per cent increase in the yield of one crop, but the need to ensure security by maintaining a diversity of life forms and particularly in plants.

Galston I certainly share your views that we are running into a very dangerous situation in terms of the available gene pools from which to draw new plants in the event of any of our present major food crops being seriously damaged by a pest which has not yet appeared. I think that we have to keep available plants which have not yet been used as food sources and also augment the research on their domestication. I think that we should not depend entirely on a few major food crops because in that direction lies possible complete disaster in the event of an epidemic of some sort.

* International Union for the Conservation of Nature, *Red Data Book*.

Gorinsky It seems to me that we should be particularly concerned about the proposition by scientists in the Hudson Institute to dam the Amazon, which is in an area with perhaps the greatest botanic speciation in the world.* The Amazon basin or the tropical East Indies are areas of extreme interest genetically and to molecular biologists, and one would have thought that there would have been an outcry, but there was stony silence.

* 'A dam across the Amazon', *Science Journal* (September 1969), *5A*, no. 3, 56–60.

Thirteen   Environmental problems and the reunification of the scientific community

Joseph G. Hancock
Department of Plant Pathology
University of California, Berkeley

Science can be misused in many ways; most dramatically and tragically for military purposes. In advanced technological nations great scientific efforts are expended on the development and promotion of frivolous material goods; narrow national and industrial research priorities circumvent world needs (colour television, space programmes). Finally, scientific discovery, applied by technology, has helped create a massive world-wide environmental crisis.

The environmental issue has generated international concern in recent years and many problems have been identified. What can science contribute towards the understanding and solution of the problems it has helped to create, and more importantly, *how can it stop creating new problems?*

Formation of a scientific community

I think that many problems of the misuse of science could be avoided if values in science were given adequate consideration in the universities where scientists are educated. In the United States the lack of social concern of many senior scientists coupled with an intellectual isolation of academic departments results in complacency about and lack of interest in the social implications of scientific research. Very little attempt is made to make students thoroughly aware of the power of science and the fact that 'impersonal' scientific discoveries create conditions that change the personal lives of millions. Consequently, few consider that scientists should be responsive to social considerations and sensitive to potential misuses. Just a real understanding of this power by its practitioners would help to create a feeling of responsibility. So far, most of our educational scientific institutions have failed even to recognize that a need exists in this area. Promising changes are beginning to take place on some American university campuses, but not nearly enough is being done fast enough. (Technological changes occur faster than science can respond!)

The manner in which American scientists regard chemical and biological warfare (CBW) research exemplifies the incredible naïvety they exhibit concerning the relationship between *their* science and *their* social responsibility. In considering CBW research, it is important to remember that much of it took place during peace-time, even before heavy American involvement in Vietnam. As is usually the case, much of the research supported by the military in universities was of questionable direct military value and was unclassified. It may even have had greater potential for beneficial than for harmful use. Under these circumstances, scientists were easily convinced of their own good intentions, and looked the other way when military policy changes made easier the use of chemical and biological warfare agents. The scientists rationalized that their research was in the national interest: that any chemical and biological warfare agents developed would only be used for defensive purposes or in retaliation. Even while members of the military questioned the development of CBW, scientists pitifully rationalized away their complicity. You could have your grant and your good conscience too! The tragic consequences of some of the CBW research are well known – herbicides and tear gas, for example, have been used in Vietnam (and with a kind of ironic justice, large quantities of tear gas have been used on US university campuses).

The problems created by defence funding obviously were not carefully considered by American scientists on a national level until after about 1966. Even now, the effectiveness of the debates over military research in American universities is uncertain. Although many university scientists are *now* feeling ambivalent about receiving financial support from the US Department of Defense, work on quasi-military projects continues in many fields.

As questioning continues about the use of science in military work, it is interesting to observe the differences in reactions of scientists in various fields. These differences appear to be related to professionalism, whereby many scientific disciplines tightly organized into professional societies become, in essence, scientific 'closed systems'. Because these groups are able to be very self-sufficient professionally, an isolationist mentality frequently develops. Broad scientific interest is often non-existent and values are ambiguous. An important effect of this scientific isolationism is that it acts to prevent the formation of a functional, responsible, general scientific community.

A dramatic example of how professionalism functions (or dysfunctions) was expressed during a symposium at the biological warfare centre (United States Army Biological Research Center) at Fort Detrick, Frederick, Maryland. The symposium, held in 1968, honoured the 25th anniversary of Fort Detrick. Symposia were centred on two subjects, 'leaf abscission' and 'entry and control of foreign nucleic acid'. It turned out, as reported in Science,[1] that many of the nucleic acid biochemists boycotted the meeting and raised a general protest against the celebration of the anniversary of Fort Detrick. Plant physiologists carried out their symposium on 'leaf abscission' apparently without concern and without mention that the defoliation programme in Vietnam was a national issue. Clearly no dramatic changes in scientists' views of their social responsibility will take place until the self-imposed isolation of various scientific groups is breached.

How to combat scientific isolationism in unifunctional government and industrial laboratories poses another problem and needs much study. Mission-oriented government and industrial research laboratories apply basic scientific discoveries. The scientists who made the basic discoveries have absolutely no control over how their work will be used. Where research is performed in the universities in return for government and industrial financial support, a form of self-regulation based upon scientific social responsibility should be imposed by international as well as national scientific communities. But, as I see it, there is now no functional scientific community and it is our job to encourage its development in every possible way.

For a community to form, many barriers will have to be removed, not the least being psychological ones. Graduate education will have to be modified to encourage the development of discipline hybridization so that younger scientists will have a greater personal feeling for the wholeness of science.

Current interdisciplinary trends in science should be encouraged. The 'environmental issue' provides a very broad disciplinary base and thus presents a promising vehicle for the task of reuniting science. American university scientists concerned with environmental problems are organizing new courses and curricula with strong multidisciplinary approaches. At the

University of California, Berkeley, implementation of these changes has forced co-operation between diverse segments of the university, and by improved communication, generated a new awareness of the social responsibility of science in improving the quality of our environment. Whether new parallel approaches will occur in research is not clear. If this should occur, it will demand changes in the current organization of American universities and scientific institutions. Fluid co-operative–consultative multidisciplinary approaches to scientific research around this single broad issue may allow an important change in the view of scientists of themselves to take place. Through involving even a minority of scientists, avenues for expanded communication will open and ideas on scientific values will be allowed to flow more efficiently through the scientific community. As a result, scientists may have a greater interest in the impact of science on society. In the long run, one of the most important effects of the 'ecology revolution' may be a reunification of the scientific community.

Research on environmental problems

Molecular biology and genetics have many implications for environmental studies. Only a few of these can be treated in this programme. I have selected two topics that I have an interest in, one on biological control and another on the biodegradation of cellulosic materials.

Interest in biological control research has expanded in recent years. In these studies, it is particularly hoped that diseases of insect and weed pests can be partially substituted for chemical pesticides, thus minimizing the need to use dangerous chemicals in agriculture where farm workers may be exposed and the environment greatly disrupted. Microbial and molecular genetics could be usefully applied in research in this field on manipulating the virulence of pathogenic micro-organisms. However, it is foreseen that basic discoveries and procedures developed during the development of so called 'super-pathogens'[2] could be exploited for military purposes in CBW. This is an example of why it is so essential that CBW research be stopped *wherever* it is done.

Biodegradation of waste products is becoming an active area of research, and one where basic studies are needed. I will consider more closely the biodegradation of cellulosic materials since these substances comprise a large proportion of the waste in some nations.

Cellulose, a large polymer of glucose residues attached by beta-1,4-linkages, is present in plant cell walls as essentially parallel aggregates that form semi-crystalline threads or fibrils. The fibrils possess an incredible tensile strength and form a mesh within which amorphous encrusting materials are deposited. The strength of the cell walls depends largely on the ordered arrangement of the cellulose fibrils. The fibrils used in the construction of textiles and paper are actually the cell walls of specialized cells from certain plants.

Plant pathogenic fungi and bacteria often cause cell wall destruction at the sites of infection. In fact, the capacity to form successful parasitic relationships in some diseases may be dependent to some degree on the ability to effect some cell wall degradation. Though the relationship between pathogenicity and the cell wall degrading ability of pathogenic micro-organisms is a complicated question, the effects of cell wall breakdown during diseases are clear. Cell walls collectively serve plants structurally as supporting elements and the integrity of infected tissues may be lost when the walls are partially degraded. Cellulose breakdown is often associated with these processes. When the heart wood of timber is attacked by wood-rotting fungi the technological value of the wood is often lost.

There has been some research on the mechanisms of cellulose degradation in the field of plant pathology, but not nearly as much as in those areas concerned with cotton and wood technology. The object of this research is, of course, to develop methods of preventing microbial attacks on cellulose, such as found in deterioration of lumber, textiles, paper, etc. The bulk of our basic knowledge on the breakdown of cellulose comes from this research.

But my concern is over the accumulation of discarded cellulosic materials. Refuse disposal is one of the major environmental problems of modern society. Cellulose in vegetable matter and manufactured plant products comprises a considerable proportion of he dry matter of garbage. In the United States over 50 per cent of refuse is paper.[3] Because of the resistance of native cellulose to biodegradation, the rate of garbage decomposition is seriously limited.

Multiple approaches towards the solution of disposal of cellulosic materials are required. Their heterogeneity makes this imperative. Studies of the feasibility of serious large scale recycling processes are appropriate. Recycling of paper in California, however, is still a volunteer activity. Publicity campaigns usefully draw attention to the problem but are totally inadequate as solutions in themselves.

One of the research approaches being considered involves accelerated biodegradation. In a sense, this is a reversal of the objectives sought by industrial scientists concerned with manufactured cellulose products. It is superimposing on existing research programmes a 'new' social need. Fortunately, the basic studies on the nature of cellulose and cellulose breakdown can be applied towards our 'new' objectives. Although these objectives involve a reconsideration and redirection of an old problem, some conflicts of interest can be anticipated, particularly economic ones. Financial support for this research should come from the industry since it is responsible for the production of the materials that are causing the problem.

As reported by Stutzenberger and his associates,[4] paper wastes are reduced about 60 per cent after composting. An anonymous note in *Nature*[5] observed that more extensive pretreatment of the raw waste might speed further breakdown of cellulose. Further use of the composted residue in mushroom culture[6] has been suggested and partially decomposed products are now commercially available as soil conditioners, mulches and plant growth media.

The fate of different papers should be considered during manufacture. Those papers that are most likely to be disposed of in large quantity (newspapers) should be constructed for rapid microbial decomposition. Studies on the relative susceptibility to microbial attack should be included in technical studies on characteristics of paper. Papers can be designed or selected for various uses partially on the basis of this quality.

As more is learned about the breakdown of cellulose, more positive manipulations of the molecular structure of wood fibres can be applied in paper manufacture so that quality can be preserved but that garbage will not.

After disposal, seeding paper with special cellulolytic microorganisms should expedite paper decomposition. Knowledge from microbial and molecular genetics should be applied to this problem. The most promising strains of thermophilic, cellulose decomposing fungi and bacteria can be selected and compost conditions adjusted, when possible, to favour breakdown of cellulose. The potential of treating paper with small quantities of materials that stimulate the production of the cellulase enzyme system (whose regulation is poorly understood) should be explored.

In addition to contributing towards the solution of plant product waste decomposition, the results of this work will contribute to related studies on cellulose preservation and plant cell wall breakdown in plant diseases. If eventually other non-biological means of dealing with paper and woody refuse are developed, this information will still be valuable.

In conclusion, I would like to reiterate that a broadly based reunification of the scientific community is necessary for 'scientific responsibility'. In hard reality, the individual scientist, expressing 'individual conscience', stands little chance of opposing today's governments and university administrations. We must assist natural processes that serve the function of reunification, such as the environment movement. But it is important also to recognize that these movements come and go and that for any lasting change we must be willing to reorganize our universities and scientific institutions, demanding a reassertion of the wholeness of science and more control over the uses of our research.

References

1 Anonymous (1968). *Science 160*, 285–7.
2 Wilson, C. L. (1969). *Ann. Rev. Phytopathology 7*, 411–34.
3 *Municipal Refuse Disposal* (1961) (prepared by Committee on Refuse Disposal, American Pulic Works Assoc., Inter-state Printers and Publishers, Inc., Danville, Illinois), p. 45.
4 Stutzenberger, F. J., Kaufman, A. J. and Lossin, R. D. (1970). *Canad. J. Microbiol. 16*, 553–60.
5 Anonymous (1970). *Nature 228*, 115.
6 Block, S. S. (1965). *Appl. Microbiol. 13*, 5–9.

Discussion

Pirani Nearly every speaker so far has talked about bourgeois problems in a bourgeois society. What I have been wondering over the past thirty-six hours is when some one is going to say something like: 'The United States is a late capitalist society and the distortions and contradictions of science which result from this can't be rectified merely by changing the structure of science and leaving society alone.'

Hancock   This is one of the reasons I spend most of my spare time in Berkeley politics. I am very depressed about the way scientists are going about changing society. Even on the Berkeley campus I haven't seen a great deal of serious political activity by scientists, though occasionally they sally forth in times of crisis. As far as actual long-term planning, actually getting down and doing grassroots work, this just hasn't appeared. However, I don't think that it's only the capitalist system which is at fault, although I have attacked the capitalist system many times in the past. I think that the socialist systems have failed in the same way. The fact is that as long as you have economic growth, as long as economic growth is considered the most important part of your society, you're going to have environmental problems. That is just a straight fact, and we've got to learn to deal with it in some way. I'm not saying that we should stop economic growth but I think we do have to tackle it and perhaps we may have to learn to adjust ourselves to an equilibrium society.

Part six Science in society

Fourteen The role of industry in applied science

A. J. Hale
Director of Research and Development,
G. D. Searle & Co. Ltd,
High Wycombe, England

I speak from the point of view of a biologist dealing with human and animal therapeutics which have had a social impact; thus I speak as one who has been exposed to the problem of applying scientific innovation in a responsible manner. For example, our company was the first to introduce oral contraceptives in 1960 and the public discussion which has centred on that topic would probably justify a meeting of its own. The pharmaceutical industry has for some forty years been the most scientifically orientated apart from the aerospace industries. Its record of achievement is considerable and so is the criticism of its attitudes and behaviour. Its background of experience in applying science is second to none and should provide guidelines for the development of responsible actions in applying science.

I worked in academic research before entering industry, so I feel at least partly qualified to make some observations on the attitudes, responsibilities and objectives of biologists. It appears to me that what we are seeking in this meeting is really a mechanism of communication. A mechanism that will not only tell the public what the scientist is doing but which will explore the implications of his work and gather and present for expert commentary the opinions formed about this work. This may lead to the formulation of mechanisms to deal with problems as they arise. I think it is too much to hope that we can predict even the kinds of problems which will arise and formulate one mechanism to deal with the lot. The mechanism created must be as flexible as the minds of the scientists with whom it will work.

There is apprehension about where science is leading us. The main question is where, but important secondary questions are who is leading us and who is apprehensive? Unless we identify those two groups we cannot set up a rapport between the two in order to influence scientific objectives. There are certain negative answers that we can give. We know that we are not being led in the formulation of what is scientifically and socially acceptable by the government, the Royal Society, any of the learned colleges or scientific societies. How then is control exercised at present? Who is apprehensive? In practice the public is the winner or loser but in general it does not care about the application of science until its immediate environment is being affected. Should we give up the Concorde for the sake of a few who might have to tolerate sonic bangs? Do you want an addiction clinic next door in order to save the youth of the country? In theory the politicians, whether local or central, should protect the population but once again it is only when a local or national crisis exists that criticism is offered. Advice is seldom sought from the scientist although it is frequently given, albeit in a highly self-opinionated way.

There is some value in presenting a comparison of responsibility in industrial and other scientific groups since it is industry which applies science.

Table 1

	Stakeholders					
	Public	Students	Other scientists	Directorate	Other funding bodies	Regulatory bodies
Universities	O	++	+++	+	++	O
Government and charitable institutes	+	O	++	++	O	O
Industry	++	O	++	++	O	++

+ equals degree of influence

Table 2

	Stakeholders					
	Public	Students	Other scientists	Directorate	Other funding bodies	Regulatory bodies
Universities	O	O	+++	O	++	O
Government and charitable institutes	O	O	++	++	O	O
Industry	++	O	++	+++	O	+++

+ equals effectiveness of mechanism of influence

The major groups of scientists which might influence the application of science are:
1. The advanced seats of learning.
2. The government and charitable research institutes.
3. Industrial research laboratories.

It is useful to examine the mechanisms by which the working scientists in these groups are controlled. The objective of a scientist is to enquire and to produce proof of the conclusions of his enquiry. To do this he requires support and an outlet for presentation of his findings. Table 1 lists the stakeholders influencing the support of scientists in these groups. One can see that the degree of influence exercised by the different stakeholders varies between the groups. Apart from the degree of influence which is exercised there is another factor which strongly modulates the degree and that is the mechanism. Table 2 shows my estimate of the effectiveness of the existing mechanisms of influence. Taken together these show that whereas there is a strong opinion-forming mechanism among scientists about their work, about not only its content but its implications, there is very little influence brought to bear on them by other stakeholders except in industry. Industrial scientists know that their objectives are influenced strongly by the public as represented by the shareholders and the public attitude to the company, the directorate, government, independent and industrial regulatory bodies and their fellow scientists. It is a difficult task to achieve scientific ends under pressures from these different stakeholders, thus I think it is worth listing the problems which beset the practical application of a scientific discovery. The list may act as a deterrent but I hope it might represent a challenge to scientists who feel that science has a practical role to play in improving social welfare. The problems I list are those that I know of in trying to introduce therapeutic products. The problems vary for each product within this category so I am certain that they will change where other biological applications are concerned. The solution pattern will however remain similar to that which has been evolved by our industry.

Scientific value and profitability

Directing an applied research and development programme is like following the horses. You have the company's money to play with and you have to decide to place your money for possible large gains on a long shot outsider and meanwhile hedge with a suitable sum on the short odds favourites. If you are efficient and have enough money you may even stick to odds on favourites and still win. Thus one's scientific decisions are always (in exciting basic work only slightly) influenced by the possible market potential of the product. Even here one can be terribly wrong in either direction. It was said in pharmaceutical circles about thirty years ago that no one would ever make any money out of steroids. I am glad that prediction was wrong.

Setting the scientific value of work is interesting. I have found that this topic is resolved by the interaction of the following groups:

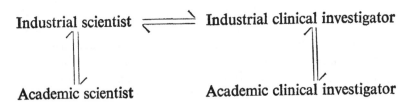

This interaction is commonplace with us and shows that we have evolved a mechanism to overcome the barrier that seems to exist in many countries between the academic scientist and his clinical counterpart.

A certain amount of work is then done to establish the efficiency of the material before the question of commercial objectives is introduced. At this point, the interaction is

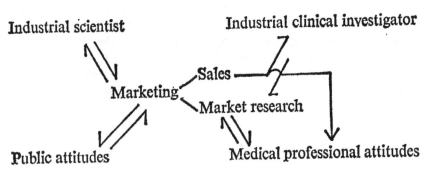

The questions which arise at this time are:
1. Can we make the product consistently, of high quality and economically?
2. Does it fill an existing need?

The role of industry in applied science 179

3. Does it create a new demand?
4. Is it acceptable scientifically to the professions that will use it?

It is astonishing how difficult it is to answer some of these questions. For example most doctors know what remedies are available but may not be too aware of the limitations of these remedies. What is their motivation to prescribe a cheaper product if the patient has to come to see them twice as often? Do they recognize that a new product may save the patient and themselves time and the nation money? Will a regulatory body recognize the value of a new, more expensive, not too different product for the sake of personal advantage to the doctor or the patient? Is the medical and lay community ready and able to accept the novelty which you introduce?

Within a company the discussions on scientific value and profitability occupy quite a large proportion of the time of many people and include the opinions of many leaders in the outside scientific community. Unless one has been exposed to the rigorous process by which a decision on expensive development of a scientific topic has been reached then one is in no position to infer that the process is inefficient. The decision is often unacceptable to some people but the mechanism used to reach it is well developed and takes into consideration the topics which I am presenting here. Decisions of this sort can never please everyone. What we have to reach in a democracy is one benefiting the majority.

Ethics

We call ourselves an ethical pharmaceutical company because our products are sold only under prescription from professional men with a high moral and ethical standard. We think that our standards are as high as theirs but now and then dilemmas do arise which are difficult to resolve.

It is not generally recognized that all pharmaceutical products carry some hazards. If a drug has an action influencing the normal or abnormal functions of the body then it will at some dosage level or under some particular functional circumstance produce undesirable effects. The problem is to find drugs which have a wide difference between therapeutic and toxic dose levels. Anyone who has taken alcohol knows that the therapeutic/toxic ratio depends on

1. The dose.
2. The formulation.
3. The functional state of the individual.
4. What other therapies are used.

How often is this appreciated in judging other drugs?

Drugs which are dangerous are frequently used as life-saving measures. In introducing a new drug where an affect on severe disease is hoped for, then risks will be accepted and taken. At what point however do you decide that the risk is not worth the benefit? Whose decision is it? Can the scientist decide before it has been tried? Does the doctor or the regulatory body decide? Is the patient ever given the opportunity to decide? Ask a patient on chronic renal dialysis if he wants a kidney transplant. The answer will be swift and definite. The problems really arise when the measure of toxic effect is very difficult to make. Is the slight thrombosis risk of the pill acceptable? Is the morbidity attributable to cigarettes acceptable? Who decides? Can you stop providing these products? Can you legislate to stop them? Could you ever not have developed them?

During the development of a product many, if not all, of these questions arise. Decisions are made to accept some of the known risks and a watch is kept for appearance of new risks. I think the industry maintains a high moral code in this respect and this is a reflection of the close interaction that takes place within it between scientist, physician and business man. None of these groups could reach ethical decisions on its own. Each has its bias and only the interaction cancels these out.

Regulatory mechanisms

These fall into three categories:
1. Internal company.
2. Industrial associative.
3. Outside governmental mandatory or advisory.

1. *Internal*

When a company has obtained positive evidence of therapeutic benefit of a potential new product under animal experimental conditions, it immediately carries out acute and sub-acute toxicity tests to establish the safety of the material. At the same time it sets up rigid specification for manufacture and quality

control of the active ingredient and of any formulations which may be used for clinical trial purposes. Also, if the scientific and commercial interest for the potential product is sufficiently high, extensive therapeutic and chronic toxicity tests will be carried out in a variety of animal species. All of this work may occupy up to five years of intensive effort. It is concerned with investigation of:
1. Natural extraction.
2. Chemical synthesis.
3. Industrial adaptation of the synthetic path.
4. Economic evaluation of the process.
5. Development of methods of process control.
6. Development of methods of quality control.
7. Pharmaceutical formulation.
8. Drug absorption.
9. Drug metabolism.
10. Therapeutic activity.
11. Acute toxicity ⎫ with chemical pathological and
12. Chronic toxicity ⎭ histopathological assessment.
13. Drug stability.

Only after all this will the company consider submitting the material for clinical trial. During this time there is constant internal criticism by the company stakeholders about the progress of the product. This is the self-regulatory mechanism developed by many companies and now adapted by government agencies for their regulatory purposes.

2. *Industrial associative*

There exists an industrial association which sets standards of commercial behaviour within the industry. Its powers are only advisory but it has the respect of the industry, the medical and veterinary professions and of the government regulatory bodies. These standards tend to be modelled on those set by the larger companies who have perhaps the greater experience of the difficulties of developing new drugs. Companies can ignore its recommendations, but this does not often happen.

3. *Outside governmental*

We have in this country an advisory committee which works closely with industry in arriving at a decision as to what is acceptable to the company, the medical profession, the government and the public in relation to the use of therapeutic agents.

When a company has completed all its own regulatory requirements and has carried out work which it knows or thinks the government regulatory body will require to provide evidence of efficacy and toxicity of a potential product, then it puts this mass of data together in a submission for approval to carry out a clinical trial. If approved, the trial will be carried out by named academic clinical investigators working in national hospitals.

This mechanism is well evolved in the US and the UK and many other countries. It is not, however, without its difficulties. If the therapeutic approach is sufficiently novel then the regulatory body may find difficulty in getting adequate independent scientific opinion to make valid judgments on the data. The danger which could arise in such a situation is that in the absence of adequate opinion the judgment might be delayed with considerable disadvantages to the company, the medical profession and the public.

To be effective, the co-operation between such a body and industry must rest on mutual recognition of the scientific standing of the persons involved. Such is the rate of change in science today that no industrial scientist can afford to wait until the basic scientific paths of progress have been defined by his academic colleagues. If he does, he will take so long in catching up with the progress of science that he will be too late in applying it to a practical result compared to his competitor. This creates difficulties for a regulatory body. If an industrial scientist is the recognized leader in his discipline then to whom will the body turn for opinion? In addition industry may have already consulted with academic leaders on the topic concerned, thus their opinion might not be considered to be unbiased. This situation can probably be resolved if such a body has powers to consult and recommend but if its powers become mandatory then difficulties of obtaining opinions to reach mandatory decisions may be difficult to come by.

Regulation can be a dangerous weapon. Premature condemnation of scientific innovation because of lack of information can be as dangerous as premature acceptance. For example – is the pill more objectionable in any scientific, medical, religious or economic sense than the Abortion Law Reform Act? A scientific comparison has never been made. Who should make it? Who should decide what is acceptable?

Commercial versus social pressure

The public does not hold shares in pharmaceutical companies merely to benefit mankind but they do retain them knowing that they are a good investment in an industry which has demonstrated that it is modern, energetic, able and responsible. Critics have considered that it is immoral to profit from the sale of therapies. Is this any less acceptable than charging for the staples of life such as food, accommodation or education? Similarly, will it be immoral to profit from other scientific advances bringing social benefit?

The finances of the industry are complex. They have been dealt with recently in detail by Happold,[1] Sainsbury,[2] Cooper[3] and Merrett Cyriax.[4] Certain factors strongly influence financial decisions related to product development. Some of these are:

1. The long period of product development. It now takes a minimum of seven years to take a potential product from its initial discovery through to clinical trial. Many products will be rejected during this progress so that it is not what it has cost to develop the scientifically successful one that matters, it is also what it cost for the failures. Even after clinical trial there is no guarantee that the product will be commercially successful.

2. It is rarely that the price of drugs rise. The prices usually remain constant and thus drop in relative value as a result of inflation.

3. The consumer is not the one who comments on prices. Neither does the doctor who prescribes. The comments usually come from government officers concerned to find economies in the national budget. This has effects on the fourth factor.

4. International awareness is such that prices in other countries tend to be pegged to the price in the country of origin irrespective of the relative costs in these different countries.

5. High risk of potential product failure. Being a leader in applied science is a very expensive business indeed. It costs around £20,000 per annum to run all the facilities and staff required to back up one PhD in a research programme. No product has a chance of reaching the market without an expenditure of at least £1,000,000.

When do you decide that you cannot afford to develop a product for which the market cannot justify the expenditure? The decision has to come before the development data can be prepared for presentation to the regulatory body. That is a decision carrying a heavy commitment and it may be fruitless if the final reaction of the regulatory body has not been properly assessed. Who will support the costs of development of a therapy for an uncommon disease? Finding a treatment for an uncommon disease may be easier than for a common one but who will pay for it? Who will make that decision where social benefit outweighs in a personal moral sense commercial return?

Patents

These were introduced to give opportunity to an inventor to have a monopoly for a certain period of time to exploit his discovery and thus to benefit from it. In stark commercial terms a strong patent position will encourage an inventor to finance the exploitation of that invention since he is safe from competition until his patent expires. He thus should be able to recoup his expenditure and make a profit. Patent life is seventeen years, so with a product development time of seven years there are ten years left to recoup the seven years of expenditure. A common misconception amongst scientists about patents is that they stop other scientists from using the material or process. They don't, they merely stop commercial exploitation of them by non-patent or licence holders. Without a patent, most companies will not proceed with development of an invention because they have no protection from piracy of their product. The surest way for any scientist to bar the application of his invention is for him not to patent it. If he feels he is exploiting the community by patenting then he is naïve about the commercial facts of life.

Strong patents mean good new products and the acquisition of international business and scientific prestige. Countries which deny patents have a miserable record of scientific innovation and are literally forced to pirate other people's products.

These are the problems which have been encountered in attempting to develop classical pharmaceutical therapies. Molecular biology has greatly increased our fund of knowledge of living things and it is certain that applications of its discoveries will be made to man and his environment in the future. From my personal point of view some of the possible applications may be as follows.

Organ transplantation

We already have fourteen officially recognized organ transplant centres in the UK and several others of equal capability who operate without official recognition. This in itself illustrates that you cannot regulate what the public wants and what individuals want to offer in the way of life-saving procedures. Organ transplants in the UK will soon rise to the level of approximately 600 a year and new forms of therapy to control transplant rejection will be introduced. Within the next two years attempts will be made to avert transplant rejection by induction of tolerance. The substances used to induce tolerance will be derived from human cells. These fall into a category of therapeutic agents which I refer to as parabiotics, since they are derived from one human cell to be applied to another and hopefully to remain in the recipient functioning in parabiosis.

Control of virus diseases

Viruses are unaffected by antibiotics since these substances kill the cells and thus the patient, if the antibiotic is used in high enough dose to kill the virus. Suggestions have been made that effective anti-viral therapies based on extraction of interferon from human cells (another parabiotic) could be introduced. Similarly modified viruses with attenuated replicating ability have been suggested by Spiegelman as short-term prophylactics or treatments for specific virus infections. At least some of these treatments will be tried within the next few years.

Correction of inherited disease

Millions of people suffer from genetically determined anaemias (thalassaemia, Mediterranean anaemias, Cooley's anaemia, sickle-cell anaemia or whatever name one uses for this group). As a result of outstanding work done by academic scientists in this country and elsewhere we now know the structure of the haemoglobin molecule and of the aberrant forms of it produced by patients suffering from these inherited anaemias. It is possible to aspirate the haemoglobin-producing cells from a patient's bone marrow and to keep the cells alive under artificial culture conditions. If these cells could be switched to produce normal haemoglobin by addition of the correct genetic coding material and then be transplanted back to the patient's bone marrow, it might be possible that the patient would develop a clone of cells producing normal haemoglobin. The biggest problems here are ones of producing enough of the correct sequence of codons and of delivering it (another parabiotic) to the recipient cell. Chemical synthesis of polynucleotide sequences is advancing rapidly as a result of the brilliant work of Khorana, so it might be possible to think of doing this kind of thing in the not too distant future. Delivery of the sequence is another matter and we have no time to speculate on that at present.

Cancer

This is the disease which most people ask about. We are learning a lot about it and we are slowly finding means of controlling certain kinds. It is not possible to see in the immediate future the emergence of a rational treatment for cancers in general. I consider, however, that by extension of genetic analysis, for example by application of the tissue typing techniques developed to determine genetic incompatibilities between organ donors and recipients as a hospital routine test, and by the development of routine immunological screening tests for cancer antigens we shall detect cancers at a much earlier stage. This could produce dramatic results in treatment of cancer patients even without introduction of new therapies. One remote therapeutic possibility is, however, related to transplant therapy. If specific cancer antigens exist then it might be possible to make them immunogenic. These immunogens (another parabiotic) could then be used to reinforce the patient's immune resistance to cancer cells (if that exists at all).

Degenerative arterial disease

For many years before your arteries harden they soften. They become infiltrated by lipid and lipid degeneration products and it is likely that hardening by fibrosis and calcification is a destructive process secondary to the infiltration. It is known that infiltration can be reversed by stringent dietary changes but in an affluent society it is impossible to educate the community to the undesirability of its dietary habits. The acquisition process is more pleasurable than the end result. Should one consider applying a therapy which will block the infiltration in the presence of dietary excess? Much money is being spent on just that sort of programme. If degenerative arterial disease can be controlled in this way, many degenerative conditions secondary to arterial change may be delayed or avoided. These diseases have a high incidence in middle-aged men of responsibility who represent a productive group in the community, and avoidance of ill health in these people is important.

At present there is no indication that a useful prophylactic will be found, so my comment here is in the nature of an appeal to concentrate on a problem which I think has a solution, rather than a statement that a solution will be found.

Industrial biological processes

Industrial chemical processes are well established and their limitations well defined. Biological processes, such as brewing, are also well established but recently more complex biological systems have appeared. The antibiotics industry is accepted as being beneficial. Production of food from biological conversion of mineral products is with us. Biological synthesis of a sweetening agent is already well advanced in our own company's laboratories,

In the future it is likely that complex polynucleotide or polypeptide products will be produced in an industrial biological process. These may be used as starting materials for other processes or as products in their own right, e.g. polypeptide hormones.

Human therapeutics have had a profound influence on social structure and attitudes during the past thirty years. Antibiotics have abolished puerperal fever, pneumonia and a host of other septic conditions as killers. Chemotherapy has closed down most of the tuberculosis sanatoria. Tranquillizers have reduced the mental institution population dramatically. The pill, where it has been available, has helped to limit population growth. We have more or less accepted these innovations for the benefits they bring but we do recognize that they also bring problems with them. Are we prepared to accept the benefits of organ transplantation, treatment of virus disease, or early detection of cancer without knowing the problems they will introduce to the community? The answer is undoubtedly yes, since the benefits cannot be denied to those needing them. We shall have to accept the problems even though we can only guess at them at present. We have a mechanism for dealing with these scientific proposals through the Committee on Safety of Drugs at present and probably in the future through the Medicines Commission. This has and should provide in the future the forum for determining what is scientifically acceptable. We know however from the trials by television and newspaper that there is either social ignorance or social unacceptability by certain sections of the community of some of these scientific decisions. For example, what are the evils of cannabis and other 'soft' drugs? We accept cigarettes, which by all present evidence are killers, yet we condemn 'pot' without knowing the facts. Should this be a problem for the Committee on Safety of Drugs? If not, who is responsible for such a decision? Parliament, the Royal College of Physicians? This dilemma illustrates our present purpose. Who should be considering, presenting, illustrating, even advertising the problems before us so that the community knows at least some of the facts before decisions are made? Such decisions must be non-political. Do we come back to the quasi-political decision-making process at a Royal Commission or do we need some other mechanism? I think that a Royal Commission is not sufficiently public in its consideration of the problem and thus has a low acceptability for its decision. We need some other mechanism. Perhaps our first function should be to answer the question, what mechanism?

References

1 Happold, F. H. (1964). *Medicine at Risk*. London, Queen Anne Press.
2 Sainsbury (Lord) (1965-7). *Report of the Committee of Enquiry into the Relationship of the Pharmaceutical Industry with the National Health Service*. London, H.M.S.O.
3 Cooper, M. H. (1960). *Prices and Profits in the Pharmaceutical Industry*. Oxford, Pergamon Press.
4 Cyriax, Merrett (1968). *Profitability in the Pharmaceutical Industry*. London, Merrett Cyriax Assoc.

Discussion

Simon You haven't considered the cost to the community of the large number of research teams working in parallel on similar subjects in various companies, and the general cost to the community of providing these very expensive therapeutic agents, though they may well pay for themselves through increased production by people who are treated.

Hale Wasteful competition is a subject on its own, as is the price of drugs. I don't think these things can be solved by Marxist principles, and I think in fact that much of what has been said of the profitability of the pharmaceutical industry can be explained; I am not saying excused – I don't think it needs an excuse.

Masters There is a widespread belief, supported by documentation in specific instances,* that much of the effort of industrial research is directed towards making new saleable products which do not necessarily have any extra medical advantages, and also that some drug companies continue to sell drugs that are profitable but may produce adverse reactions.

Hale I would say that we, and I'm talking about major companies with major responsibilities, go to great lengths to collect information on adverse reactions, and in fact probably have more data on them than the regulatory bodies. Regulatory bodies tend to act in a national manner – we act in an international manner; in fact we bring international data together to present to a national body.

Rosenhead Our one speaker from industry has talked about 'the role of industry in applied science'. Now this is not the title in the provisional programme, which read 'Is industry scientifically irresponsible?' Nor is it the title we might have expected, which is 'Is industry *socially* irresponsible?' Industry has an enormous effect upon the community. It has a direct effect upon science through its selective application of funds for particular ends; all that Dr Hale has said about this is that industry increases the freedom of academics by offering them alternative sources of funds.

Hale I said that the academic has an alternative source of funds which gives him independence within his organization for carrying out his free ideas.

* See Consumers' Union of United States Inc. (1970). *Consumer Reports* 35, 616.

Rosenhead *That* is the only thing Dr Hale said about the effect of industrial funds on the direction of science. He didn't say anything at all about the funding of contract research, by which funds are allocated to particular areas because of their industrial applicability rather than their social value. He didn't say anything about the allocation of priorities within industrial research establishments, nor about industry's sensitivity to social concerns, nor about the problems which industry produces for society, both through effects on the environment and by their concentrations of political power. He didn't talk at all, or very little, about the possible social irresponsibility of pharmaceutical companies, except to say that he opposes a mandatory regulatory system. All these issues of industry and society are far more complex than the questions he talked about today – which were concerned with mechanisms, mainly internal to industrial firms. We should have ample opportunity, preferably with Dr Hale present, to discuss these issues.

Hale I think I'm speaking for industry as a whole when I say we recognize these difficulties and that we should like to be involved, and I think we should be involved, in their solution.

Watson I'm afraid my reaction to Dr Hale's presentation was that it was largely irrelevant, almost completely avoiding the question of how best to serve the general public with drugs. The major problem facing us now is not the difficulty of getting new drugs on the market, but the correct way to ensure that the general public receives the best drug treatment compatible with the limited amount of money society has to spend.

Hale If I make a sweeping statement about research institutes or universities, and I damn them all for the sake of one or a few who have particular attitudes, that is unfair, grossly unfair. In the same way, if you make that criticism about a large industry, perhaps because of the activities of a few members, that is equally unfair.

Monod This whole business of the drug industry is, and ought to be, an extremely sensitive one. On the other hand, it is an extremely complicated problem, and I think it's entirely unjust to put it all on Dr Hale's shoulders. He has given us a perfectly reasonable and satisfactory picture of how he goes about his work. I know something of the drug industry because in a way I'm in it myself as a member of the Pasteur Institute and in fact a member of the Board of Trustees; the Pasteur Institute is a drug company, if you like, with one difference: all its profits, and it makes fairly nice profits, go entirely to research, including research that is completely non-directive and academic in character. This is a very fortunate situation and I cannot help feeling that in a reasonably organized society this is gradually what should happen.

Perutz Isn't this also true of the Wellcome Trust?

Monod Yes.

Beckwith I agree with most of the comments made to Dr Hale but I wonder why people have been so polite with other speakers who I believe have been equally irresponsible in facing the implications of their work. Dr Hale is pretty much alone as a representative of industry and at this sort of meeting we very rarely have any contact with people from industry whom we have essentially ostracized. I think one of the most important needs is to build the ties between academic and industrial scientists.

Dauman I wonder how many criticis of Dr Hale have ever been in industry, or seen industry, and where they get their information from. Surely we are not talking about irresponsible industry or irresponsible science, but irresponsible people, and, there's just as much irresponsibility, as Dr Beckwith was saying, within the academic scientific community. This irresponsibility never seems to get criticized within that community with quite the same sort of emotion as that in the criticisms of Dr Hale and the pharmaceutical industry.

Fifteen   Science and social values*
Martin M. Kaplan
Director, Office of Science and Technology
World Health Organization, Geneva

In this paper I shall examine certain aspects of the application of systems analysis to society's problems. My reason for doing so is that in recent years the systems approach has come to be considered by many as a possible planning instrument for dealing more scientifically with the intricate web of social and scientific problems that beset our world. Since scientific method, by its very nature, requires objectivity, we can only welcome its increasing integration into the analysis and solution of these questions.

However, this approach to social problems presents formidable difficulties, and not only technical ones. The ideological aspect is infinitely more complex, and has to do with the definition and evaluation of the myriad social factors involved and ranging them on a scale of priorities. This is the essential groundwork on which solutions can be constructed.

The question then arises as to who should be responsible for setting social values and goals. To me it seems obvious and necessary that this process should involve all segments of society; that, in fact, these issues should be thrashed out by the public at large. In spite of the difficulties inherent in pursuing such a path, it is imperative that we do so, and ways must be found.

I make no claim to being an expert in general systems theory or in systems analysis as developed in the last thirty years. Some of my qualifications I share with other non-specialists, including Monsieur Jourdain in Molière's *Le Bourgeois Gentilhomme* who found he had been using prose all his life without being aware of it; another part I share with some of my colleagues who have been struggling, until now without great success, to apply such techniques to medical and public health problems.

* The views expressed in this article are the author's own.

Science and social values

Systems analysis is a way of achieving a defined objective by considering alternative possibilities, and then selecting and arranging those that promise optimum efficiency in a highly complex network of interactions. It makes use of both mathematical and verbal models in defining its elements and objectives, and requires the application of many disciplines, as well as the use of computers for processing the great bulk of material involved.

This approach, sometimes called cybernetics or operational research, has helped in coping with certain moderately complex issues. The striking successes of the British group of scientists with operational research during World War II, and the rapid development of computers after the war have resulted in increasing research and application of this and associated techniques in the military and industrial fields. These techniques are now being brought to bear on major areas of science, technology and society. However, the problems involved in fields such as industry and the military are relatively simple as compared with social issues. As the complexity of the problem increases, the difficulties in applying the systems approach increase many-fold. Nevertheless, this method is now applied in such areas as town planning and other aspects of urbanization, and environmental degradation and its psychosocial and ecological effects. The pace of application to all sorts of problems is swift, and decisions are being made quickly.

But it must be kept in mind that some of these decisions may be wrong or even dangerous if the decision-makers are beguiled by the numbers game. Schlesinger,[1] for example, notes that the systematic use of quantitative analysis has solved certain cost-benefit problems in the United States military forces. However, he shows that by allowing the economic or other single low-order factors to outweigh political, social and other 'soft' or intangible elements, serious mistakes have sometimes been made. He cites instances of high-order strategies in NATO and the war in Vietnam in which bureaucratic inertia and faulty estimation of an opponent's response have operated importantly. The latter element, in his view, represents the greatest danger to realistic decision-making – for instance in predictions of the USSR's response to the ABM, and North Vietnam's reaction to its economic and casualty losses. Schlesinger goes on to say: 'systems analysis has provided something of a facade of pure objectivity for political debate. While insiders know better, systems analysis has been presented to the public as an instrument that somehow "solves" problems. The upshot has been to obscure the unavoidable role of political imponderables in decision-making and to discourage analysts from dealing explicitly with these imponderables.'

How then does one treat these and other imponderables in a systems approach? To begin with, these components are set up as verbal models. The systems analyst then attempts where possible to express them quantitatively for use in computers. In general it is preferable to use a verbal model than an oversimplified mathematical model. Even when it cannot be formulated quantitatively, a model in ordinary language has value as a guide, and it is moreover open to challenge. Such a systems approach to social problems should have a flexible base providing for redefinition of values and priorities. Thus, the greatest possible freedom would be left for future planners.

Another systems specialist, Ozbekhan,[2] observes that long-range planning involving social systems is highly complex, full of contradictions and paradoxes. He notes, for instance, that these systems must be broken down into parts for operational purposes, rather than pursued as whole systems. Also, decisions indicated for the present may prove bad for the future, but he suggests that the ill effects of any particular decision could be minimized by buffering with other components in a plan. He goes on to say that in the absence of any integrative general theory, quantitative analysis and prediction of an exact nature is not possible in long-range planning. A positive point Ozbekhan makes is the following: in the systems approach as applied to social planning, an early setting of clearly formulated goals will exert great influence on the synthesis and implementation of the plan. He terms this in effect a 'willed' future.

The concept of a willed future has precedents throughout history. The promulgators of the Magna Carta, national constitutions, bills of human rights and bodies of law all set out to define their goals as clearly as possible. In so doing they created in some degree the conditions they sought. The new element in systems analysis is that it offers the closest approximation to a scientific method for achieving desired social aims.

Let us now turn to a set of problems that will illustrate some of the considerations mentioned above: planning for delivery of health services. What, for instance, are the criteria that should be used to allocate finite medical resources to individuals and groups in a given population? Should more emphasis be given to birth control than to the saving of lives; to the provision of hospitals and clinics for the present population rather than to the increased training of medical personnel to meet the needs of the future? Should priority be given to malnutrition in the young rather than in working age groups; to the problems of cancer and cardiovascular disease or to improved water supplies; to preventive medicine in general rather than to curative medicine? The allocation of a large part of a health budget to malaria eradication, for instance, means the sacrifice of attention to many other health problems in a developing country with meagre resources. Or, the construction of expensive hospitals may mean the neglect of sectors of preventive medicine that could benefit greater numbers of people.

Answers to such questions depend inevitably on the relative values assigned to various possible choices by decision-makers. While such decisions in the health field are now made with some attention to their inter-effects, individual problems are often treated in isolated fashion with little reference to other areas of social and economic development. A systems approach that would attempt to identify and link all these factors would obviously be superior to present practices. This would require the frequent use of verbal models in the first instance followed by the assignment of values to various alternatives in constructing a priority scale. Such a procedure has at the very least the virtue of revealing neglected components of the problem and of crystallizing ideas in an otherwise vague conceptual framework. Also, in the field of health we have found that if problems are approached in an unrelated way, e.g. by individual sectors such as communicable diseases, mental health, degenerative diseases, serious imbalances and errors will result. Such faults might be overcome by first delineating over-all health and other social and economic objectives, and then working downwards through a systems network to individual components. Such attempts are only just beginning; to date we can boast of little success – and that little is confined to narrow areas of operation.

The overriding needs of developing countries – food, medical care, education – are immediate and urgent, and leaders in these countries cannot be expected to be greatly concerned with broad and long-range considerations. Nevertheless, I believe that if the planners in these countries were to acquire some grasp of the principles involved – some awareness of the usefulness and the limitations of the systems approach – they would be less likely to commit serious errors.

Despite the difficulties and the notable lack of success to date in applying systems analysis to social problems, it would I believe be short-sighted and irresponsible to disregard or neglect this area of endeavour in the world today. It is true that many countries at present can be said to have 'planned' societies, and some of them employ systems techniques to a limited degree. And yet we are faced with war, poverty, increasing disorder and social alienation, distorted priorities, declining freedom and individual powerlessness. These are products not of man's inherent evil, but of the inexorable grinding of the national machines with their imperatives of growth, profit and glory.

The crux of the matter, then, is the necessity to determine what our social values and goals are, or should be. If not universal, can they be made compatible with the social goals of other groups so as to avoid serious conflict? Is it ultimately possible to define, weigh and choose between the social values implicit in such moot concepts as democracy, socialism, capitalism, national sovereignty, freedom, progress and so forth? All too evident are the enormous difficulties of arriving at a consensus on such questions, and in measuring the imponderables of politics and of individual and group reactions. Moreover it would be naïve to expect social action to conform closely to a blueprint.

These are serious and perhaps insuperable impediments to the application of a systems approach to broad social issues. But there would appear to be no alternative. It is not possible to remain content with a patchwork approach to peace, widespread poverty, and the other great social problems. Even partial victories are not to be scorned in these perilous times, and the systems approach properly applied offers, I believe, the soundest available method for achieving certain limited social goals without seriously compromising the objectives of a willed future.

As stated earlier, the participation of the general population is essential in helping to shape values and to make choices. Here the scientist is on a level with other citizens. However, scientists have a special role in systems techniques, and this confers a special responsibility.

Accordingly, I would propose that as scientists we participate more actively in such efforts; and, more precisely, that we specify values and goals in operational terms as far as possible. Also, in our search for answers to problems we should give due regard to their long-range aspects and interactions, and identify them in a scale of value priorities.

The use of a systems approach is a road spiked with uncertainties. If narrowly conceived and applied it could even become a Frankenstein's monster. If, moreover, a scientifically ordered world brings us, as Bertalanffy suggests,[3] a choice of Aldous Huxley's *Brave New World* at best or George Orwell's *1984* at worst, it is not too happy a prospect.

But, as scientists who share a faith in rational thought applied to a disordered world, we must exploit whatever analytical and synthetic tools we have, and try to perfect them. The systems method, whatever its limitations, cannot be ignored and will undoubtedly be developed and used to ever greater degree. Whether for good or ill will depend largely on ourselves.

References

1 Schlesinger, James R. (1968). *Bull. atomic Scientists*, November 1968, 12.
2 Ozbekhan, Hasan (1968), in *Perspectives of Planning*. Paris, Organization for Economic Co-operation and Development.
3 von Bertalanffy, Ludwig (1968). *General System Theory*. New York, Brazillier.

Discussion

Solet The systems approach offers a number of dangers. Scientists especially should realize that this is a business approach, developed essentially by economists, where one is assigning values to various courses of action. It's an economic way of thinking, even when it doesn't involve dollars. The danger with a systems approach is that it tends to favour those objectives which are most easily definable, most mission-oriented and most likely to succeed, at the expense of projects which offer much less obvious and immediate pay-offs. The kind of research which Dr Watson mentioned, the kind which very few would undertake because of the low possibility of success, becomes a very poor risk under a systems approach.

The systems analysis approach is also a threat to democracy

because it does nothing but glorify the cult of the expert. As it gets more and more complex the possibility of lay members of society contributing to decisions which influence the course of the analysis becomes increasingly remote.

Kaplan I'm not thinking of social values in terms of dollars or pounds; I'm thinking of them in terms of what society itself thinks is important, and in putting a weight on them – and such decisions are not in the hands of the élite. The major point that I wanted to make is that this is the responsibility of the public, and the scientist is in no better position to put these social values into a weight framework than any other citizen.

Sixteen Evolutionary biology and ideology: Then and now

Robert M. Young
Wellcome Senior Research Fellow in the
History of Biomedical Sciences
King's College, Cambridge

This paper is concerned with 'the attempt to develop a more adequate intellectual framework' for understanding 'the general principles involved in relating science and society'.[1] In particular, I want to consider the relationship between science on the one hand and philosophical, social and political problems on the other. I hope to provide some suggestions which will help us to see the constitutive role of evaluative concepts in biology and will leave us to discuss values and politics as such: not cloaked in the specious objectivity of ideologically neutral positive science. I shall begin by contrasting this point of view with what I take to be the usual piecemeal approach to the study of science and society. Next I shall suggest that the philosophical status of certain key concepts in biology relate them as closely to the human and social sciences as they do to the physico-chemical ones. This point makes the introduction of the concept of ideology in biology much less contentious than it might appear to be at first sight. For reasons which I shall outline, the discussion will be conducted for the most part at one remove from molecular genetics and will concentrate on the general theory on which all modern biology is based. The examples which I will discuss are drawn from the 19th-century evolutionary debate, from Lysenkoism, and from current social and political extrapolations based on evolutionary biology. I hope to show that these form part of a continuous tradition in which it is routinely impossible to distinguish hard science from its economic and political context and from the generalizations – which often also serve as motives for the research – which are fed back into the social and political debate. The conclusion which I hope to support is that we will have to learn to think in new ways if we are really serious about exercising social responsibility *in* science.[1]

I hope that it will not be taken amiss if I begin by saying that it seems to me that most discussions of social responsibility in science reflect a Fabian approach to the fundamental problem, that is, favouring a gradual, piecemeal strategy rather than attacking the enemy head-on. I should perhaps remind you that the concept of Fabianism is double-edged: the resulting tactics, in the hands of Fabius Maximus (surnamed Cunctator, the delayer) did wear out the strength of Hannibal, while at the same time they were based on the fact that Fabius lacked the resources to meet Hannibal in open battle. It is clear that we lack the intellectual and moral resources to attack on all fronts, but it is surely worthwhile to begin to enquire about the equipment we would need.

The result of our piecemeal strategy is that we persistently formulate our problems in terms of science on the one hand and values on the other, although there are tentative attempts to define their relationship. Looking at the papers which have been presented at this conference, one finds that they consider aspects of technology, industry, medicine, the very mixed fruits of the applications of science in the form of genetic manipulation, immunology, agriculture and the environment. Finally, we have been told about fragmentation and about the direct political role of scientists. If we add to these the crisis in the funding of science, the dramatic consequences of linking immunology with surgery, the side effects of chemotherapy, and the very direct effects of defoliants – followed in some contexts by other sorts of very grotesque fragmentation – one is left with a picture of the ship of scientific objectivity buffeted by the winds of the military-industrial complex, technology, medicine and so on. On such a buffeted ship, how do we who are also morally concerned, responsible men conceive and carry out our sense of social responsibility? About all we can do in these circumstances is to gasp out our complaints: 'We caught you.' 'You can't do that to me (or my findings).' 'I won't do it.' 'I'm concerned about that.' Some move on to ask, 'What are we going to do about it?' and go on to shout, 'Stop that!' or 'Get on with this', while a few drop out to do something which they find more morally satisfying or socially relevant. This last group is related to a mounting (and more or less coherent) critique of the scientific world view and its relations with the ethos of advanced technocratic societies.[2] Much of the strength of the counter-culture and the appeal of pop pseudo-biology stems from the failure of professional scientists to ask certain questions in relevant ways.

At this point we come up against the fundamental assumptions of modern science and find that we are the victims of our own myths. The central problem lies at the heart of the view of science which we hold and propagate. We are struggling to integrate science and values at the same time that we are prevented from doing so by our most basic assumptions. It would be ludicrous to attempt briefly to discuss the metaphysical foundations of modern science, but it may be useful to remind ourselves of certain key issues and to mention some concepts which bear directly on scientific explanation in biology.

In the 17th century the development of methodology and of the quantitative handling of data was related to a fundamental metaphysical shift in the definition of a scientific explanation. The concepts of purpose and value – the 'final causes' and teleological explanations – which had been central to the Aristotelian view of nature, were banished from the explanations of science (though not from the philosophy of nature). The questions one asks of nature could be as evaluative and qualitative as one liked, but the answers had to be made in terms of matter, motion and number. In the physico-chemical sciences this list of so-called 'primary qualities' has been modified to include some less precise concepts such as force, energy and field, but the fundamental paradigm of explanation – the goal of all science – has been to reduce or explain all phenomena in physico-chemical terms. The history of science is routinely described as a progressive approximation to this goal. This is the metaphysical and methodological explanation for the fact that molecular biology is the queen of the biological sciences and the basis on which other biological (including human) sciences seek, ultimately, to rest their arguments. I need hardly say that this has been a rather forlorn goal for much of biology and the source of a great deal of punning and sheer bluff.

The task of demonstrating the role of ideology in the most nearly physico-chemical aspects of biology is, in principle, the same as that of providing an ideological critique of the fundamental paradigm of all post-17th-century science. This task has been undertaken by Whitehead, Mannheim, Burtt and others, and an assessment of it cannot be made here. In leaving this question aside, however, we should not let the undoubted success of molecular biology obscure the fact that most of biology is far from qualifying for the more difficult task of requiring a metaphysical critique. For the most part the biological sciences lie half way along a continuum extending from pure mathematics and the physico-chemical sciences at one end and the woolliest of the human and social sciences at the other. The particular consequence of this intermediate position which is most unpalatable is that biology partakes as much of the philosophical and methodological problems of the social and political sciences as it does of the physico-chemical ones. One can support this argument by pointing out that there is a hierarchy of concepts in modern science which extends from the purely physico-chemical to the purely evaluative and that biology shares a number of the most significant ones with the 'softest' sciences.

At the fundamental level one finds the *primary qualities* mentioned above, and these are employed to explain the subjective or *secondary qualities* of colour, odour, taste, temperature etc. In biology these qualities are the terms in which we analyse *biological properties* such as irritability, contractility and so on. (The concept of a 'biological property' was a conscious departure from the official paradigm of explanation, and continues to serve us well.) Properties are the terms in which we analyse *structures and functions*, and in doing so we employ (along with the human and social sciences, which cling obstinately to organic analogies) the concepts of *adaptation* and *utility*. Structures and functions are the terms in which we analyse the next level of explanation, *organisms*. On the basis of the theory of organic evolution, biologists argue, of course, that *persons* are organisms, but the concept of a person retains a further analysis from an older metaphysical tradition and continues to be subjected to a dualistic division of the *mental* and the *bodily*.[3] I want to return for the purposes of this argument to the concepts of structure and function and trace some of the related concepts along a different path. It takes only a moment's reflection to see that the related concepts of adaptation and maladaptation, normal and pathological, health and disease, clean and dirty,[4] adjustment and deviance[5] are very relative indeed, and the employment of them is seldom far from explicit or implicit moral (and often political) values. It is now a commonplace of the philosophy of science that all facts are theory-laden.[6] In biology, many facts are related to concepts which are inescapably value-laden, and the same concepts are used – sometimes directly, sometimes analogously – in the human and the socio-political sciences. By this point it should not be thought too great a jump to introduce the concept of ideology.

The term 'ideology' has traditionally had derogatory and political connotations which are connected with its popularization by Marx, who concentrated his use of it as a term of abuse for ideas which served as weapons for social interests. But Marxists were soon subjected to their own critique, and this led to a general definition of ideology: 'when a particular definition of reality comes to be attached to a concrete power interest, it may be called an ideology'. Before Marx, however, those who coined the term and who called themselves *Idéologues* considered themselves to be straightforward scientists who argued that 'we must subject the ideas of science to the science of ideas'. Their efforts in epistemology, psychology and physiology helped to lay the foundations for modern experimental medicine in France, but Napoleon found that the *Idéologues* were opposing his imperial ambitions, and his criticisms and oppressive activities gave the term a derogatory connotation. Recent writers have attempted to re-establish a value-neutral use of the concept in the discipline of the sociology of knowledge.

The connection between what I was saying about the position and concepts of biology along a continuum, with that of ideology, should become clear if we adopt the point of view of the sociology of knowledge which argues that situationally detached knowledge is a special case and that situationally conditioned knowledge is the norm. Knowledge is both a product of social change and a factor in social change and/or the opposition to it. This is a commonplace, but its systematic application has radical consequences for the idea of 'objective' science. The fundamental claim is that our conception of reality itself is socially constructed. You will recognize the essential insight in Marx's oft-quoted assertion: 'It is not the consciousness of men which determines their existence but, on the contrary, their social existence which determines their consciousness.' More recently it has been argued that no human thought, with the exception of mathematics and parts of the natural sciences, is immune from the ideologizing influences of its social context. It is in this sense that the sociology of knowledge offers itself as a tool for analysing the 'social construction of reality'. If we adopt this point of view, we can approach the problem of the relationship between science and society from a new perspective. Although the sociology of knowledge was developed as a result of problems in the social sciences, it can be argued that our own problems should lead us to apply it to similar questions in natural science and especially in biology. Just as the concept of a hard, discrete fact has had to be given up in the philosophy of science and the pure sensation in psychology, the scientific concept which depends on these — that of 'objectivity' — must surely be brought under scrutiny. Going further, the privileged place of science in society and culture, sharply cutting off its substantive statements from values, politics and ideology, must surely be examined very closely.

I appreciate that the point of view which I am advocating is itself ideological, but it is not purely so. At the same time I am arguing that we should search for these factors — not, I hasten to add, to expunge them but to discuss social and political issues as such — I would equally argue that the case for the role of such factors depends on presenting evidence which convinces a morally concerned and critically thinking man. The point is that there is no escaping the political debate, a debate which extends to the definition of ideology but also to that of science and its most basic assumptions.

In its early manifestations the concept of ideology conveyed a sense of more or less conscious distortion bordering on deliberate lies. I do not mean to imply this. Like the concepts of alienation and exploitation, ideology does not depend on the conscious intentions or the awareness of men. Nice men exploit, and contented men are alienated, just as honest men have false consciousness. To deny this would be to commit the intentional fallacy, a polemical device which is widespread enough these days. I know a professional manager of vast estates who claims resolutely that his work has nothing to do with politics, while at the opposite extreme Angela Davis and the American Black Panthers claim that all black people in prison are political prisoners. Similarly, just ten years ago Daniel Bell proclaimed *The End of Ideology*. Unfortunately, the book in which he did this contains fulsome thanks to organizations, publications and individuals who have since been shown to have close financial and political links with the American CIA.[7] Thus, the effort to absorb the ideological point of view into positive science only illustrates the ubiquitousness of ideology in intellectual life.

Having spent most of the available time in outlining the philosophical issues involved in the effort to relate biological science with values, I can only sketch some of the evidence which I believe justifies the use of ideological analyses in biological problems. I shall mention three case studies which were chosen because they raise the issues starkly and have been examined in sufficient detail so that one can safely refer to the secondary literature: the nineteenth-century evolutionary debate, Lysenkoism, and the current trend of writing speculative politics in the form of pseudo-biology.

Most historical research on the development of the theory of organic evolution has stressed one of two themes: (1) the scientific story based on geology, palaeontology, zoogeography, embryology and domestication, along with the post-Darwinian debate on the validity of the mechanism of natural selection, leading eventually to the neo-Darwinian theory with its basis in genetics and molecular biology.[8] (2) The other perspective is the Victorian debate on the conflict between science and theology which eventually centred on evolution.[9] But there is a third and equally important theme in the whole story, one which contributed to and derived from the scientific and theological issues. I want to use this aspect of the debate as the basis for the analogies I shall make about the recent past and the present.

If one both broadens and narrows one's perspective on the nineteenth-century evolutionary debate, it emerges that social and ideological factors defined the context of the debate at the same time as they determined key issues about the narrowest scientific problem: the precise mechanism of evolutionary change. This context involved a number of complexly interrelated issues which cannot be considered here: natural theology, Utilitarianism, phrenology, historiography, belief in progress, positivism, and so on. If we follow the thread of the scientific debate, it leads from the economic writings of Adam Smith and T. R. Malthus to the theological and ethical works of Paley, to the theological geology of William Buckland and Adam Sedgwick, to the equally theological – but anti-literalist and anti-evolutionary – writings of Charles Lyell, and on to Darwin, Spencer and Wallace. This debate was closely intertwined with and fed directly into controversies in psychology, physiology, medicine, sociology, anthropology and genetics, all of which were invoked in debates on 'Social Darwinism' and imperialism. There is not at any point any clear line of demarcation between pure science, generalizations based on it, and the related theological, social, political and ideological issues.

However, if one were forced to choose one issue which was more nearly central than any other to the whole debate, it would be the role of *struggle* in defining the relations between men and between man and his environment. Was the competitive struggle for existence inevitable, inescapable, and even ordained, and did it or did it not produce moral and social progress? The Malthusian theory of population provided Darwin with the key to the central analogy between changes produced by the selective efforts of the breeders of domesticated animals and the process of natural selection. Although a great deal of controversy about the meaning of Darwin's theory for man and society was conducted in his name, Darwin resolutely declined to take part in it. One's analysis of the role of ideology in his work lies, therefore, in the context, the genesis, and the debate into which his ideas fed. But even Darwin pointed out that every fact must be for or against some theory. He might have added that for practically everyone else, facts and theories were exquisitely relevant to social, political and ideological positions in the Victorian debate.

Alfred Russel Wallace, the co-discoverer of the theory of evolution by natural selection, was also indebted to Malthus for his insight into the mechanism of evolution. However, he very soon saw that the basis of the mechanism in Malthusian theory came into direct conflict both with his socialism and his philosophy of nature. Consequently, he abandoned natural selection as applied to crucial issues in man's physical, mental and social development. He drew explicitly on anti-Malthusian social theories in doing so. He concluded (rightly) that Malthusianism was used by conservative and liberal thinkers as an excuse for blaming nature for man's inhumanity to man and taking a fatalistic view about the impossibility of radically restructuring society.

Herbert Spencer, on the other hand, was actively seeking a mechanism which would guarantee social progress, and he saw that the Malthusian analogy could not provide that. We tend to think of Spencer as a Victorian prig and a champion of the losing side – the Lamarckian theory of the inheritance of acquired characteristics. In doing so we forget two important historical facts. First, the question of the mechanism of evolutionary change was wide open throughout the nineteenth century (even Darwin became progressively Lamarckian in his thinking) and was not resolved in favour of neo-Darwinism until well into the 20th century. Second, Spencer was very influential in nineteenth-century biology, and his social theories were far more influential than those of Darwin and Co.: so-called 'Social Darwinism' is a misnomer.[10] Spencer is quite explicit about the role of ideology in his view of the mechanism of evolution. He had turned to biology to find support for an extreme version of individualist *laissez-faire* social theory (vestiges of which have been evident at this conference), and he thought he had found it in Lamarckianism. Towards the end of his life he prefaced his umpteenth defence of the inheritance of acquired characteristics (in a debate with Weismann) with the following remarks: 'a right answer to the question whether acquired characters are or are not inherited, underlies right beliefs, not only in Biology and Psychology, but also in Education, Ethics, and Politics'.

The definitive answer to Spencer's hope of evolving the perfect society, if only men would stop interfering with inevitable progress by ill-considered things like public health measures, state schools, a postal system etc., came from T. H. Huxley. Between 1860 when Huxley smote Bishop Wilberforce's theological pretensions against Darwin's theory, and 1893, when (again at Oxford) he delivered his cautionary lecture, 'Evolution and Ethics', his defence of biology had moved from casting aside a simplistic theological account of life to earnestly advocating that men realize that science and evolutionary theory could not provide a guarantee of progress or a substitute for moral and political discourse. In the meantime evolution had been invoked to support all sorts of political and ideological positions from the most reactionary to the most progressive, from total *laissez-faire* to revolutionary Marxism. The fallacy which Huxley was combating was the naturalistic one. While agreeing that we cannot infer human morals, much less inevitable social progress, from science, we should not fail to see the complementary point that moral and political views were already deeply imbedded *in* the science of the day.

I hope that I have made plausible the claim that the nineteenth-century debate was far from free of ideology at any level. The link between this debate and the notorious case of Lysenkoism was one of the would-be participants in the evolutionary debate. Karl Marx wanted to dedicate the English edition of *Das Kapital* to Darwin, who politely declined and wrote to a friend, 'What a foolish idea seems to prevail in Germany on the connection between Socialism and Evolution through Natural Selection.' Marx wrote to Engels that he saw in Darwin's theory 'the basis in natural history for our view'. However, Marx and Engels were at pains to divorce evolution from its Malthusian basis. Like Wallace (on the Left) they saw that the Malthus–Darwin view of natural selection was available as the basis for a reactionary, fatalistic view of man's position and social change; and like Spencer (on the Right) they saw that it failed to provide a guarantee of social progress towards Utopia. Marx and Engels thus rejected natural selection and tended to support Lamarckianism as more congenial with their view of nature, history and social change. Once again, one should recall that the experimental evidence was at that time open to a number of theoretical interpretations.

There are those who would argue that since about 1900 there has been a decisive shift in evolutionary biology and that the progressive working out of genetic and molecular mechanisms has brought major aspects of its fundamental research into such close contact with pure physico-chemical science that the role of social, political and ideological assumptions is rapidly becoming vanishingly small. But in every period since the Renaissance it has been claimed that the level of positive science has finally been reached in biology. It was thought that the paradigm of modern biology was firmly established in the midst of the 19th-century debate, and it is thought again now. The point is that *at the time* it was impossible clearly to separate the factors. We are now in an analogous period when people are debating the social meaning of biology, but before mentioning aspects of the current debate, I want to touch on Lysenkoism.

There is little point in my reviewing the controversy: there is an excellent book and some other good studies of it.[11] However, I would like to suggest how we might approach the literature on Lysenkoism. Even the term evokes in us horrors of the suppression of a scientific tradition, censorship, pure ideological invective at the expense of objectivity in science and at the expense of agricultural yields in a hard-pressed country. It also conjures up the awful consequences of the cult of the individual bolstered up by Western anti-communist pressures extending from the Revolution through the Cold War. It should be recalled that it was Stalin's direct support for Lysenko which was decisive. This continued under Khrushchev, and the catastrophe in agriculture which was partly attributable to Lysenkoism played a role in Khrushchev's downfall. Western scientists see the Lysenko episode as pure, rank abuse of science and use it to shore up the anti-communism which they acquire from other influences. In sum, not very relevant for us.

But the fine texture of the controversy is very illuminating just *because* we are so complacent about it. I suggest that we attempt to study it in a different light, not as pure distortion or pornography but as the sort of pathological exaggeration which we find so useful in biological research in illuminating the norm. As Professor Gombrich points out, caricature can reveal important features by means of grotesque exaggeration.

Two contrasting points will suggest what I mean. First, from the point of view of at least quasi-objective experimental science, the controversy is very reassuring. The men who stood out against all the hardships of the period had something to cling to – the methods and findings of international genetics and agrobiology. The lengths to which the Lysenkoists had to go in expunging all traces of chromosomal biology is an inversion – a reversing mirror – of the rational structure of biological science, showing the way it hangs together as a network of evidence and inferences. They had to rewrite textbooks in every field of biology, medicine, psychology, pedagogy and so on, and to institute censorship at every stage of publication. The contrasting point is that from an ideological point of view Lysenkoism makes perfect sense if we see it in the light of the continuing controversy leading from the 19th-century debate. As one of the Lysenkoist enthusiasts wrote, 'Weismannism–Morganism serves today in the arsenal of contemporary imperialism as a means for providing a "scientific base" for its reactionary politics.' Another said, 'It disarms practice and orients man towards resignation to the allegedly eternal laws of nature, towards passivity, towards an aimless search for hidden treasure and expectation of lucky accidents.' A typical article in the period was entitled 'Mendelist–Morganist genetics in defence of Malthusianism'. In the same period J. D. Bernal wrote a withering critique of a number of neo-Malthusian socially pessimistic works by eminent British scientists – including a Presidential Address to the British Association – which were based on just the analogies which the Russian writers mention.[12]

It is rarely the case that the history of science produces such a clear-cut example of the attempt of ideology to root out the well-attested findings of a rapidly-developing research tradition, culminating in a conclusive physico-chemical explanation of the basic mechanisms involved. It seems to me that this episode provides a very promising research laboratory for studying the limits of the ideological analysis of biology. Two conclusions are already clear: it is seldom the case that one is dealing with pure science or pure ideology. Multiple causation is the rule.

Second, a whole generation of biologists in Russia learned to see nature in Lysenkoist terms and to do science in good faith within that framework. I said a long way back that concepts of health and disease, adjustment and deviance are very relative indeed. It is worth remarking that Medvedev was committed for a time to a mental hospital for having allowed his book on the Lysenko affair to be published in the West. In an important sense he was mad to do it, but strong protests led to his release instead of committing the protesters as well.

This leads me to the current debate. Professor Bettleheim tells us that student radicals are suffering from neurosis (many, he assures us, can be cured by psychoanalytic psychotherapy).[13] Professor Lorenz explains human aggression and student protest in ethological terms (less hope there), while Herbert Marcuse tells us that the positions of both the young radicals and the old reactionaries are biologically determined.[14] The list of authors who have recently written ideologically prescriptive works in the guise of descriptive and generalized accounts based on genetics, ethology, archaeology and anthropology, and general biology is by now familiar to most of us: Morris, Ardrey, Comfort, Towers and Lewis, Koestler. It is growing daily. From the point of view of professional scientists, one can feel safely distanced from this pseudo-biology. We do not take it seriously when Lysenko cites pseudo-evidence against intraspecific competition, or when Robert Ardrey claims that masses of scientific data support the inevitability of such competition. We can even feel that it is little to do with us if Professor Darlington – with 'FRS' prominently printed on the bookjacket – cites a mass of scientific and historical evidence punningly interpreted in support of reactionary social doctrines, including *apartheid*.[15] (We knew he thought such things, after all.) But just as the Lysenkoists argued that modern genetics gave support to Western bourgeois reactionaries, it is clear that Professor Darlington's pseudo-science gives comfort to the South Africans. For example, when a highly critical review appeared over my name in the *New Statesman*, I had a letter from a South African graduate patiently explaining that my reading of the book was a result of my ideological bias. My point is that of course he was absolutely right about me, *and* I am right about Professor Darlington's book.

More and more people are trying to base generalizations about man, society, culture and politics on the biological sciences. They have always done so and will continue to do so. Many of them may be relatively easy targets, but the essential point is that no one can confidently draw the line between fact, interpretation, hypothesis, and speculation (which may itself be fruitful). It seems to me that it is the social responsibility of science to enter wholeheartedly into this debate and directly answer such works in the non-specialist press. Paradoxically, we must relax the authority of science and see it in an ideological perspective in order to get nearer to the will-o'-the-wisp of objectivity. We have won a Pyrrhic victory in establishing the part-reality and part-myth of the autonomy and objectivity of science, and the existence of this Society and its conflicting aims reflects our unsteady position. In one sense science should feel strong enough to stop flailing horses which died in the nineteenth century in their attempts to protect the status and methods of science. But in another sense, we need – for our own moral purposes – to think seriously about the metaphysics of science, about the philosophy of nature, of man and of society, and especially about the ideological assumptions which underlie, constrain and are fed by science. Since we have systematically weeded out this tradition among working scientists – one which flourished until the 1920s – we need help from other disciplines in gaining the necessary perspective, and we could well turn to the continuing traditions of enquiry in the social and political sciences which have gained impetus from the civil and international conflicts of advanced technological societies.

We can, if we are reluctant to consider evaluative concepts as integral to biology, retain the distinction between the scientific and the evaluative. Although I believe that the maintenance of this distinction is philosophically indefensible, the programme which can be recommended for the further study of social responsibility in science is operationally indistinguishable from the one which follows from the strong version of my thesis: study works on ideology and social science and apply their analyses to our own work in order to test the limits of pure science. I suspect that little will remain inviolable, but whether or not I am right about this, we will have cultivated a perspective which encourages the evaluative and political consideration of scientific concepts and will find ourselves in greater control of extrapolations from our work and much more wary of the specious aura of scientific objectivity in which they are cloaked. I am in no sense recommending an anti-rational, much less an irrational, activity. The aim is to open up more aspects of science and its context to public debate so that conflicting values can be discussed *as such*. Science is no substitute for morality or politics, nor is it independent at any level from them. We need to see that ideology is an inescapable level of discourse and need (in the first instance) to debate conflicting ideological positions and (in the last instance) to face and resolve the actual conflicts between the needs and goals of men in the appropriate way.

References

1 For readers who would prefer a short list of books on these issues, the following are recommended: A. N. Whitehead, *Science and the Modern World* (Cambridge, 1925, also paperback); E. A. Burtt, *The Metaphysical Foundations of Modern Physical Science*, 2nd ed. (London: Routledge, 1932; also N.Y., Anchor paperback); K. Mannheim, *Ideology and Utopia. An Introduction to the Sociology of Knowledge*, trans. Wirth and Shils (London, Routledge, 1954; also paperback); P. L. Berger and T. Luckmann *The Social Construction of Reality. A Treatise in the Sociology of Knowledge* (N.Y., Doubleday, 1966; also Anchor paperback); C. W. Mills, *The Sociological Imagination* (N.Y., Oxford, 1959; also Penguin).
2 E.g. Roszak, T. (1970). *The Making of a Counter Culture. Reflections on the Technocratic Society and Its Youthful Opposition.* London, Faber.
3 E.g. Strawson, P. F. (1959). *Individuals. An Essay in Descriptive Metaphysics.* London, Methuen. Part I.
4 Douglas, M. (1966). *Purity and Danger. An Analysis of Concepts of Pollution and Taboo.* London, Routledge; also Penguin.
5 E.g. Ingleby, D. (1970). 'Ideology and the human sciences', *Human Context* 2, 425-54.
6 Feyerabend, P. K. (1962). 'Explanation, reduction, and empiricism', in Feigl, H. and Maxwell, G. (eds), *Minnesota Studies in the Philosophy of Science, Vol. III: Scientific Explanation, Space, and Time.* Minneapolis: Minnesota, pp. 28–97; Scheffler, I. (1967). *Science and Subjectivity.* N.Y., Bobbs-Merrill.
7 Bell, D. (1962). *The End of Ideology. On the Exhaustion of Political Ideas in the Fifties*, revised ed. London, Collier-Macmillan; also Free Press paperback.
8 See Wilkie, J. S. (1959). 'Buffon, Lamarck and Darwin: the originality of Darwin's Theory of Evolution', in Bell, P. R. (ed.), *Darwin's Biological Work. Some Aspects Reconsidered.* Cambridge, also N.Y., Wiley paperback, 262–307; Eiseley, L. (1959). *Darwin's Century. Evolution and the Men who Discovered It.* London, Gollancz; also N.Y., Anchor paperback.

9 E.g. Greene, J. C. (1959). *The Death of Adam. Evolution and Its Impact on Western Thought.* Ames, Iowa; also Mentor paperback.
10 E.g. Hofstadter, R. (1955). *Social Darwinism in American Thought,* revised ed. Boston, Beacon; also paperback.
11 Medvedev, Z. A. (1969). *The Rise and Fall of T. D. Lysenko,* trans. Lerner. N.Y. and London, Columbia; Mikulak, M. W. (1970). 'Darwinism, Soviet genetics, and Marxist-Leninism', *Journal of the History of Ideas 31,* 359–76.
12 Bernal, J. D. (1952–3). 'The abdication of science', *Modern Quarterly 8,* 44–50.
13 Bettleheim, B. (1969). 'Obsolete youth. Towards a psychology of adolescent rebellion', *Encounter 33* (September), 29–42.
14 Marcuse, H. (1955). *Eros and Civilization. A Philosophical Inquiry into Freud.* Boston, Beacon; also N.Y., Vintage paperback; London, Allen Lane, 1969; also Sphere paperback.
15 Darlington, C. D. (1969). *The Evolution of Man and Society.* London, Allen & Unwin.

Discussion

Edge The scientist, in fact, is very much in need of the kind of critical self-awareness which Bob Young has just provided: and historical analysis like this is a prominent feature of the science studies courses which we provide for science students at Edinburgh.* But there are other relevant perspectives. One which I think adds greatly to the self-awareness of scientists, and therefore to the aim of making them more rational, more explicit, more autonomous, more effective (which I take to be the function of the debate in which we are engaged here), is the perspective which comes to us through social anthropology.

Young One of the things I was trying to convey is that we are in a position where we really need deliberately to take a rather odd perspective, and I think the anthropological one is very useful indeed. I feared it might be very difficult to make that palatable to an audience of this kind, but I'm very glad that you brought it up because it seems to me to be the next step in trying to gain some self-awareness.

* For further details of these courses, reading lists, etc., write to The Secretary, Science Studies Unit, Edinburgh University, 34 Buccleuch Place, Edinburgh EH8 9JT.

Bohm You pointed out that ideology has entered into all sorts of questions which are mixed up very deeply in science, in the whole way science is done, and that this could have led to a lot of bias, and that in fact it did, for example, during the Lysenko affair. But there seems to be implicit in your point of view, that although you admit you have ideology, you yourself are able to look at this ideology in some unbiased way. If there is bias, it means that there is something wrong with what we say, something false. You see a bias means that you accept something false in order to support a position that is dear to you. Perhaps we could say that in addition to social sources of bias there are individual sources of bias, peculiar to each individual; in other words, the ultimate ideology is the self of the individual.

So I think that this raises a deeper issue, which I think has not been considered, which is that a bias distorts true perception and must therefore be destructive. The question is, what are we going to do about such bias, if we say we are aware of it?

Young If I have understood you correctly, we disagree. And one of the things we disagree about is the relationship between bias and objectivity. You think that men can arrive at a consensus; I think I live in a world where the goals of men conflict so dramatically that consensus is a very long way away. And as long as we live in that kind of world then the question of bias, which of course is a methodologically important question, is to my mind completely swamped by all sorts of things to do with repression and misery and genocide.

An extensively annotated version of this paper appears in *Science Studies* (1971), vol. 1, no. 2, containing references to numerous related issues, publications and evidence supporting the argument given here.

Seventeen The myth of the neutrality of science*

Steven Rose and Hilary Rose
Professor of Biology, The Open University;
Lecturer in Sociology, London School of Economics

'Science is neutral' – the history of a debate

A recent analysis of the social responsibility of the scientist by Nobel Laureate Ernst Chain italicized the statement that 'science, as long as it limits itself to the descriptive study of the laws of nature, has no moral or ethical quality and this applies to the physical as well as the biological sciences.'[1] It is this set of beliefs and ideas concerning the neutrality of science which has begun to wear the aspect of a myth, which while presently ubiquitous in socialist and capitalist, industrialized and non-industrialized societies, none the less is of relatively recent origin. Our purpose here is to challenge this myth, for we argue that the acceptance of it has been to a large degree responsible both for the anti-human applications of science and to its current moral crisis.

The 19th-century conflicts over the role and relationships of science and the limitation of the concept of pure science and society have been discussed by Bob Young. Here we will show them at work in 20th-century science as well. Although one may regard Engels, in his *Dialectics of Nature*, as having provided a preview of much of this analysis, it was not until ten years after the revolution in the Soviet Union that a specific attempt was made to come to terms with the concept of pure science. This period was typified by attempts to define a 'socialist physics', 'socialist biology', and so forth. The form of such a socialist science was uncertain and open to debate. What was clear was that it ought to be different from 'capitalist' science, and that, in this scheme of things, there was no such thing as 'pure' science.

'Pure' science was simultaneously under ideological attack, not

* An extended version of this paper is to appear in *Impact of Science on Society*, vol. 21, no. 2, 1971.

only from the left, but the right, for the emergent Nazism of Germany was beginning to talk of 'Aryan' and 'non-Aryan' (i.e. Jewish) science. Some branches of physics, notably Einsteinian relativity and quantum theory, were under attack for their non-Aryan quality, and there were cases when the same type of physics was attacked both for its non-Aryanism, and for its non-socialist, 'idealist' nature.

The reaction to these attacks upon its self-perceived integrity by the majority of the scientific community was predictable. In Germany, and the Soviet Union, the response was one which Haberer[2] has characterized as the politics of 'prudential acquiescence'. This acquiescence led in Germany to the dismissal of Jewish university scientists and the official acceptance of an openly racialist biology, justifying the organized and monstrous slaughter of the concentration camps. In the Soviet Union, it took the milder form of prudential acquiescence in the systematic destruction of a particular field of science, genetics, and the exile or silencing of its champions.

In pre-war Britain, the debate took a different turn. Traditional empiricism reduced a discussion of whether the internal logic of science itself was ideologically determined to one concerning the harnessing of science to human welfare. Reformist and Marxist scientists generated a wave of optimism about the prospects and significance of science as a factor in the liberation of mankind. To achieve this effect, science must be rationally planned and organized. In an appropriate social structure, science would inevitably proceed so as to enhance human welfare. Even this proposition, however, resulted in much of the academic community closing ranks against the onslaught, proclaiming its self-interested purity with virginal fastidiousness.

War, and the mobilization of the scientific community in the prosecution of war, made the debate meaningless. Even the most prudential of the German scientists found themselves involved in the war effort – some Jewish scientists even managed to survive the war in their laboratories. In Britain, the purest of academics were registered and drafted into war research. And in the USA, the biggest scientific mobilization the world had ever seen, the Manhattan Project, took place. The Republic of Science became a dream of peace which science, even when the peace at length came, was never really to recover.

The era of Big Science was ushered in by the explosions at Hiroshima and Nagasaki, and the debates of the 1930s seemed

The myth of the neutrality of science

strangely irrelevant. Science was paid for by government; the largest part of this payment was for war (defence) science, but governmental and industrial research contracts permeated the universities too; particularly, of course, in the USA. But, surprisingly, the neutrality myth has not withered as Big Science has emerged. The reasons for this seem to relate back in part to the debate of the 1930s and 1940s, which were effectively terminated for many scientists by the Lysenko episode. The revelation of the distortions introduced by the imposition of Lysenkoism in Soviet biology threw Marxist Western scientists into a state of intellectual disarray. Some left the Communist Party, like Haldane. Most retreated from an outspoken defence of the prospects of a socialist science into a more neutralist position.

A second, and perhaps still more important, factor in the withering of the neutrality debate was the dilemma faced by many physicists who had taken part, many with the noblest of intentions, in the Manhatten Project. Believing that Hitler was developing the bomb, and that Hitler must be stopped and Fascism destroyed, what choice was there *but* to attempt to provide a bomb for the Allies? The fact that Hitler was *not* on the way to getting a bomb, and that the ones that the physicists had made were used on Hiroshima and Nagasaki, provoked a crisis of conscience. If their profession was to be saved, and their consciences salved, they needed to discriminate between the *effects* of this use of their physics, and the physics itself. The physics had to be neutral; only the use to which it is put need be condemned. And as more and more it became apparent that, in the West, science was being applied to evil ends, the need to maintain the distinction between the subject and its use became sharper. For precisely those who in the past had argued that the link between science and human progress was inevitable, the retreat back to the laboratory and its neutrality became a necessity *if they were not to stop doing science altogether.*

An interesting parallel evolution seems to have taken place in the Soviet Union. Again, it may be related both to the failure of Lysenkoism and to the success of the bomb. It is exemplified by the changing Soviet view on nuclear weapons. In the early 1950s the official view still reflected a belief that technology could be made to serve Man's ends. The bomb was seen as an adjunct to, but not a transformation of, class war, which by definition must be victorious. The language of nuclear holocaust, the Doomsday machine, and the jargon of the US arms control experts was repugnant. But by 1970, this view was no longer held in the Soviet Union. The bomb is now seen as transforming war.[3] The jargon of the two super-powers is a mutual one, the language of the SALT talks. Technology is seen to be just as sweetly inevitable to the Soviets as it was to Oppenheimer, when asked his views on the US H-bomb. And the corollary of technological inevitability is neutrality. 'Science', a distinguished Soviet physicist assured one of us recently, 'is neutral; it is how one uses it which determines its good or evil potential.'

It is only in the last few years that this ubiquitous acceptance of a science simultaneously closely articulated into the bureaucratic, military and industrial machine of contemporary society, and yet freed from responsibility, has been effectively challenged, though the challenge has tended to take the form of the rejection of *all* science as irrevocably linked with the instruments of state oppression; the relationship perceived by Marcuse between the products of contemporary physics and the needs of IBM and the US Atomic Energy Commission, has, we have argued elsewhere[4] resulted in a specific rejection, not only of the sciences of the establishment, but of the whole methodological apparatus and rationality of science itself. It is this rejection which brings the conflict right into the heart of science in one move.

The philosophical issues

One way of illustrating the non-neutral nature of science is to examine the constraints which operate on science within the present system. Thus, if we recognize that Big Science is state-financed, and that there is always more possible science than actual science, more ideas about what to do than men or money to do them, the debate is, in a sense, short-circuited. Science policy means making choices about what science is to do. It is not a question simply that science is inevitable and cannot be stopped, science is always being stopped and started – by withdrawal or injection of funds. Whoever makes these choices about what to finance, by definition they cannot be ideology- or value-free; they imply an acceptance of certain directions for science, and not others; opening certain routes means closing others. Putting a man on the moon means not doing other sorts of things. Such choices are inherent in any system. And as they are clearly not neutral choices, the science they generate cannot be neutral.

The myth of the neutrality of science 219

It is this fact, of the reality of social control of what science has done, which makes at best redundant, at worst smug, some of the moralizing over the social responsibility of the biologist which has been at times apparent in the dramatic over-writing of some science journalists. Shall we clone people, or make test-tube babies? What about sex determination, or genetic engineering? Should the socially responsible scientist carry the moral dilemma of all these issues, a humanitarian demigod in a white lab coat? The truth is that these are specious questions, rhetorical non-issues which distract attention from the real problems which face us. For all these developments, if they are to come, are regulated not by the wish-think of the men in the labs, but by the fact that they are expensive programmes of research and development which will have to be funded by some agency or another on the basis of particular criteria. If the funds are forthcoming, no amount of posing moral dilemmas will stop *some* scientists doing the work, even though individuals might opt out. Witness CBW, for example. Monod may be right; 90 per cent of scientists may refuse to work on war research. But 10 per cent of scientists may be more than enough.

The debate then is about power; about social control over science. The problems of science relate in part to the fact that this power is not shared; decisions are private and not open – the responsibility of the scientist should be to ensure that this power is shared with the people at large, rather than to be seen staggering under self-made moral burdens which look suspiciously like bundles of feathers.

These statements clearly relate to the objective constraints on present-day science. But the constraints we have so far defined are external ones. They are not quite adequate for our purpose for they still do not approach the core of the neutrality question. Thus we have asserted that the social environment itself is enough to specify the direction of research and that no research can be divorced from its environment. But there remains the problem of the present responsibility of contemporary researchers for future developments which they may not be able to predict. Consider the following cases, for example.

When an American woman scientist was awarded the US Army's highest civilian award, for the development of a more effective form of rice-blast fungus, specifically suited to Southeast Asian conditions, this award was clearly not given for neutral science. But what about the basic science that led up to this work, which may not have been done under military contract or in a defence establishment? Take the development of the tear-gas CS, now extensively used in Vietnam, for example. This gas was developed in the mid-1950s by Britain's chemical defence establishment at Porton Down, as a result of a recognition of the inadequacies of the then used tear gas, CN, on a number of technical criteria. Researchers at Porton began their search for a new agent as a result of a specific directive from the British Ministry of Defence. In the course of screening a number of possible agents they came across CS. Bulk production followed. Thus the work was done for a specific objective; it was clearly not value-free; by definition not neutral. Porton work was plainly mission-oriented. As a mission cannot be neutral, the science done in achieving it cannot either. But what about the work from which the Porton studies were derived? Are we to indict Corson and Stoughton, back in the 1920s, for the initial observation that ortho-chlorobenzylidenemalonitrile was a lachrymator, just because thirty years later Porton picked it up and used it for a new purpose? If we were to do so, we would also have to include all the other hundreds or thousands of academic researchers doing their 'pure' research in the laboratories of the 1920s, churning out their three papers a year on the properties of odd chemicals, and the behaviour of model systems, simply because they were working in the ambience of a society whose structure imposed a consequence not in harmony with human welfare. We will have not merely to indict Rutherford and Einstein for the atomic bomb, but practically all the chemists, physicists, mathematicians and biologists who have published research in the present century. Plainly, this is a *reductio ad absurdum.*

It might then be argued that there is a cut-off point in the neutrality debate. Non-mission-oriented, basic research whose immediate application is not apparent might seem excluded from it. At a pragmatic level, such a common-sense idea might seem acceptable, provided the research was not sponsored by a fund-giving agency or organization with a mission other than the support of basic science for its own sake, such as Departments of Defence or industry. Particle physics and molecular biology might seem to come into this category.

Whilst such a view might appear sound enough as a rule-of-thumb for everyday practical purposes, it avoids recognition of the possible interconnections between science as a cognitive system and the social system. That is, it assumes that, whatever

The myth of the neutrality of science 221

the goal choices made by funding agencies, within those financial constraints the actual content of the science which is done depends only on the objective accretion of data, facts and theories. It thus implies a purely 'internalist' view of the nature of scientific knowledge, as a set of ever-advancing and self-consistent absolute approaches to a statement of 'truth' about the universe. This view of science, though, is one that has come under serious challenge recently from the philosophers of science. Thus the activity of scientists can be divided, according to the illuminating insight of Kuhn, into 'normal' science and 'revolutionary' science.[5] Normal science is what most scientists do all of the time, and all scientists do most of the time; it is solving a set of puzzles about the natural world. The puzzles are designed and solved in terms of a paradigm, a gestalt view of the world, which provides a framework for normal science. At certain periods in science, and for a variety of only partially understood reasons, Kuhn argues, there occurs a paradigm switch, a change in the gestalt view of the world amongst working scientists, which alters the puzzles they set themselves. This paradigm-switching is what Kuhn calls revolutionary science: problem-solving instead of puzzle-solving.

Corson and Stoughton's work is essentially of the puzzle-solving kind. What we wish to propose here is that, while it is not possible to ascribe a 'value', an element of non-neutrality, to *all* single pieces of puzzle-solving basic science of this type, it *is* possible to ascribe a value to the paradigm within which they are conducted. Puzzle-solving basic science of itself, unlike mission-oriented science or development, cannot have either neutrality or non-neutrality ascribed to it; the concepts are irrelevant. They are relevant only to the paradigm within which the puzzle-solving activity is being conducted.

But a paradigm is never value free. A paradigm is never neutral. Hence while we do *not* have to search for non-neutrality or its moral congener, responsibility, in the work of a particular puzzle-solving scientist, we find it without difficulty in the paradigm within which he is working. Some 19th-century examples of this effect within biology have been provided by Young. But the 20th-century metaphors of cell biology and biochemistry are both different and at least as interesting. Anyone reared on the biochemistry that developed from the 1930s on will recall that the central theme of teaching and research was that of *energy*. The key to the workings of the cell – referred to as the cellular *economy* – was the flow of energy within the cell and the clue to this energy flow was provided by a substance known as ATP. It was Lipman in the early 1940s who produced the key metaphor for the activity of ATP; it was the energy *currency* of the cell. Oxidation of glucose resulted in the synthesis of ATP – storing it in an 'energy bank'. A compound related to ATP, creatine phosphate or CP, is synthesized when there is an abundance of ATP; it is referred to as an energy 'deposit-account' compound compared with the ATP 'current-account'. Energy, economy, banks, current and deposit accounts... such terms are more than mere puns; they both reflect the vision that the cell biologist has of his phenomena and help to direct his thoughts to new experiments. Much of the biochemistry of the 1930s and 1940s through to the mid-1950s was conducted within this type of language system, with its pre-Keynesian undertones.

From the mid-fifties on, though, the language system began to change. The central metaphor was altered; the paradigm switched. Many of those responsible for this paradigm switch have contributed to the present meeting; rarely has such a galaxy of metaphor-makers been assembled in one place. What are today's central metaphors? We have already heard them used in the last couple of days. They are those of control, community and communication, feedback, interaction, repression and regulation, switching on and switching off.

These are the metaphors of cybernetics and the engineering and computing sciences; but still more they are the metaphors of today's managerial, post-Keynesian capitalism or bureaucratic socialism. They form the scaffolding for today's puzzle-solving research in biology. And in their turn they serve to bolster the structure of society by providing it – by virtue of a measure of determined extrapolation – with a biological rationale as well. The role of such metaphors cannot simply be subsumed as punning; they frame too closely the thought processes of the researcher.

The metaphor, the paradigm, sets the questions we ask of our subject, and hence the answers we seek from our materials. If a behaviour geneticist asks the question: 'How much does heredity determine intelligence?', he has limited the answers to his question before he has begun his research – that is, he has located his answer within a particular paradigm. It is the question and its framework, not so much the answer, which are non-neutral, their historical antecedents belonging to the line of eugenics stretching back to Galton and beyond. And we should not be surprised to find this type of research providing certain types of answers, which are then clearly related to certain social and political purposes – as has been the case with the work of Jensen in the US, for example. It is not that the question should not be asked, but that we need to be very clear about the nature of the paradigm which sponsors it and which, within the limits of the data, specifies the answer, as it must for all puzzle-solving research. For much of science, the analysis of possible non-neutral components in specific scientific paradigms is very difficult. It may well be that it is only in periods of social and intellectual crisis that we can glimpse the interconnections between science and the social system.

To suggest the non-neutrality of science is, we realize, to recognize the passing of a myth which has nurtured academic science and its considerable achievements for more than a century. In doing so, however, we are not assaulting either the *objectivity* of science done within its paradigm, nor are we decrying the validity of science's inner logic. The advance of science, despite the shifting perspective of its paradigms, presents successively more accurate approximations to comprehensive statements about the nature of the universe. The way in which men see the world may vary with their viewpoint, but the variation will be confined within the crucial limitation that, for all of us, our viewpoint is human; there is a finite specification to the way in which our brains function, and to the relationship of this functioning to the environment.

Hence, in so far as objectivity specifies a public viewpoint that is more or less common to all men, there is an objective internal logic to science which new techniques, experimentations and paradigms help refine. The universe may be (can be?) solved in this sense, granted time – and a paradigm which does not precipitate disaster. But the language in which the solution will be framed will depend upon the paradigm; of the infinity of questions we can ask about the universe, we will choose some of them and not others from within the ideological framework set by the paradigm.

When, as they now are, issues of human survival are on the agenda, it means that the time has come when it is right to assert, as a necessary counter-myth, that it is possible to use the techniques and methods of science for man-centred and not man-destroying purposes; that we must build human relevance into the paradigms of science itself.

References

1 Chain, E. B. (1970). 'The responsibility of the scientist', *New Scientist*, October.
2 Haberer, J. (1969). *Politics and the Community of Science*. New York, Van Nostrand-Reinhold.
3 Moss, N. (1970). *Men who Play God*. London, Penguin.
4 Rose, H. and Rose, S. (1969). *Science and Society*. London, Allen Lane The Penguin Press.
5 Kuhn, T. S. (1970). *The Structure of Scientific Revolutions*. Chicago University Press.

Discussion

Sussman I don't believe in objectivity in science so I want to quarrel with you about how one places social values on what scientists do. You proposed a rule based on Kuhn's work where the social value comes out of the paradigm and not the activity of puzzle-solving. I'll try to put this into a realistic framework. If two people are taking genes out of *E. coli* – one person taking the lac gene in order to study the regulation of lactose transcription, and the other to get a gene out to make a very toxic bacteria grow like *E. coli* – I think they are both performing within the same scientific paradigm and yet it's their puzzle-solving which is different. In other words, it gets back to what they want to do with their science which puts the value on it rather than the scientific paradigm of being able to take genetic material out of a bacterium and introduce it into another bacterium which is the same in both cases.

H. Rose I don't see that raises much of a problem because it seems to me that you yourself have indicated they would have rather different motivation, so at that level of just ordinary human goodness or wickedness the case stands. What you have said is that a particular bit of problem-solving may be used by a bad man or a good man and it will have different outcomes. What I am saying is that the values themselves are built into the paradigm. One can see this very clearly, I think, in the case of 19th-century biology.

Eighteen The scientist in opposition in the United States

Jon Beckwith
Department of Bacteriology and Immunology
Harvard Medical School

I wish to speak about problems that are facing many young scientists in the United States. The causes, the concerns, are also seen in other countries. However, they are more keenly felt in the United States because of the increasing awareness of the role of our country as a major force of oppression and exploitation in the world and the awareness of the part that science plays in these activities. This level of concern in the United States is not a new phenomenon; the atomic physicists faced similar problems and confronted the same crises many years ago. Unfortunately, because of our training, and, in particular, the way in which science is taught in the United States, we are having to learn the same lessons all over again. I hope that these lessons will be learned in time, and, more important, that we will go beyond the tactics of the atomic scientists in confronting the social impact of our work.

I should like to begin by talking about the experiences of a group of us who last year attempted to make a political statement on the occasion of our purification of a bacterial gene. Although what I will discuss reflects the opinions of a few, I believe the motivation for our step derives from the same contradictions which are affecting many other young scientists.

The group of us involved in this action had developed strong feelings about the way in which science is used in the United States. We had seen research, sometimes deriving from basic bacterial genetics, used to create weapons of biological warfare. We had seen drug companies exploiting antibiotic research to enrich a few and exploit the many. We had seen a misdirection of health priorities – where super research is carried out on heart transplants and cancer and little attention paid to major health problems such as malnutrition, lead poisoning, infant mortality and delivery of health care in general. Of course, the use of overwhelming technology to destroy the people and countryside of Vietnam played a role in our growing concern. We tried to point out some of the contradictions confronting scientists working in the United States under these conditions. As an example, the progress in molecular genetics (of which our work was a graphic illustration) was clearly leading many into research in genetic engineering. There are now a considerable number of scientists, many leaving work with bacteria, who are attempting the first simple experiments in this field. We cannot predict when the first such experiments will be successful or when the techniques will be applied. However, we cannot fool ourselves into believing that the applications are far enough off so that we bear no responsibility for their implications. A group of prominent scientists, in conjunction with a private foundation, have proposed a governmental crash programme of funding for genetic 'disease management'. It was suggested to me by one of the founders of this group that the programme be analogous to the Manhattan Project!

I don't want to overemphasize the dangers of genetic engineering. There are certainly many other dangers already present or probably nearer in time in terms of their applicability. However, it is certainly a question molecular geneticists should be discussing. Furthermore, I do not believe that the directions and applications of this work should be decided by a group of 'prominent scientists' any more than it should be decided by a group of politicians. There are political questions involved that require a much wider participation in decision-making – a participation that our present system does not allow.

We concluded that unless the present system in the United States is changed, there will always be the danger and the likelihood of the use of science for the exploitation and oppression of people all over the world. We don't pretend that this problem is peculiar to the United States or to capitalist countries; but at least a prerequisite for a change to a society where there can be much greater participation in decision-making is the elimination of a system which is based on the exploitation of man by man. We felt that scientists bear an important responsibility in alerting people to these dangers and in ensuring control over the use of science.

The reactions to our statements were strong and bitter from some quarters. There was an outcry from many scientists against publicizing any negative aspects of our work. They felt that the problems and control of science were better handled quietly by leaders of the scientific community. These viewpoints are reflected in comments by the editors of *Nature*. 'What justification can there be for supposing ... the perversion of an unknown technique in the hands of a medical profession which, for all its faults, has consistently worked in a beneficient direction ... The progress of science itself may be interrupted or even halted by excessive fear of the consequences.'[1] I must respond that if the speed with which science is progressing must be slowed down in order to spread the benefits of science among all people, so be it. However, this is not a necessary consequence of a re-examination of the role of science. If science is affected adversely by public fears, I believe that the inaction of scientists themselves will be, in large part, responsible for this negative result. So many have become scientists because they felt that they could contribute to the betterment of mankind, and so few have done anything to ensure that science was in fact used in this way.

People who with us spoke of the dangers of uncontrolled science were called 'doomsday men' and 'jeremiahs'.[2] We make no predictions. One does not have to predict doom to a Vietnamese or to a thalidomide baby. We were recognizing our own complicity in a destructive system.

These then were our concerns and the reaction of a fearful scientific élite. The debate over these issues heightened with us the contradictions we faced as biologists working in the United States. How do we then justify our continuing to do such research or do we decide to change? This same question is being asked by many young scientists. What paths are open to students who have begun careers in science?

First of all, there are those who feel that they should change to some area of science or medicine where there is a possibility for directly benefiting people. It should be remembered, however, that much research which can be directly beneficial to people can also have harmful results. *There is no escape from the burden to the scientist.* The scientists still have no control over the way their discoveries are used by government or industry. The medical researcher cannot ensure that some new drug or technique he finds will be distributed fairly without regard to income or that a drug company will not exploit it to increase its profits. Certainly he can do beneficial research, but whom will it benefit? Some idealistic medical students have become disillusioned about the possibility of their contributing to an improvement in society.

Some of these quit for the above reasons and also because even if they were to do medicine in areas where people are most deprived, they are only serving to patch up a problem which ultimately requires a political solution.

So, there are some who are dismayed at the negative applications of science, who see no positive way of contributing as scientists, and therefore quit science altogether. I believe it is an illusion, however, to think that there is much possibility for finding some other role in our society which does not in one way or another serve the goals of the few who run our government and its institutions. *You cannot escape by quitting science.* To earn a living in the United States it is difficult not to be complicit and supportive of the system. For instance, many students have moved from the natural sciences into social sciences. Yet, one example should suffice to show that workers in this area confront the same problems as we done. One of the largest federal investments in the social sciences has been the $8 million 'Cambridge Project', a joint Harvard-MIT programme funded by the US Department of Defense. The purpose of this project is to stimulate the extensive use of computers in the social sciences. The proposal for the project, which is public, includes a section on the potential wide range of uses to the military of such a project. These uses include data banks and analyses of radical movements in the US, computerization of interviews with Viet Cong prisoners, computer analysis of the ways to best reach peasant populations in order to understand how to make them more patriotic, etc.[3]

I cannot deny that there may be some roles we might play in our society in which it is possible to contribute much more than the scientist to movements which will change society. However, choices are obviously limited by an individual's talents and interests. It would probably be difficult to make a radical historian out of every scientist with radical political views.

What then should be the tasks of the radical scientist who sees that he cannot contribute to radical change in any major way through his science? He is unlikely to escape the contradictions by quitting science. I believe there are some very important tasks.

1. In the same way that radical historians or economists expose the way in which most history or economics is taught from a political viewpoint supportive of the system, the radical scientist must expose the way in which science in the US is politically organized and directed. We are educated to believe that progress in science is one of the most important factors in improving the

welfare of mankind. We must respond that science can and has been used equally for retarding the progress of man towards a better society. In addition, as Seikevitz says, 'It seems to me to be disastrous to think that further refined technology, directed though it may be, will cure past technological mistakes ... technology has badly outraced the political and social means of handling the problems it generates.'[4]

The view of science as a purely progressive force, the lure of prizes and many other factors justify the most corrupting levels of competition in the more favoured fields of biology, for instance. The organization of labs for the greatest efficiency leads to degrading master–slave relationships between supervisor, student and technician. What benefit is a cure for cancer to man, if in the process he is losing his humanity? All of these features must be exposed as part of a radical analysis of American science.

2. There is a very limited role for the radical scientist in contributing his scientific knowledge to those groups who are working for radical change. For example, doctors, nurses and technicians are lending their expertise to free health clinics being set up all over the United States by such groups as the Black Panthers and the Young Lords. These clinics are being established as models for the way in which health care would be delivered in a changed society. Electronic engineers are contributing their knowledge to provide the Black Panthers with protective electronic equipment. Such organizations as the Medical Committee for Human Rights and Science for the People work on these principles. However, it appears to me that only a very limited number of scientific fields can provide such useful talent. The scientist who follows this path may often face great risks in his career; but groups such as the Panthers are facing far greater risks.

3. The scientist must recognize his responsibility for control over the applications of scientific research. However, this control cannot and should not be exerted by scientists alone. First, scientists working as a group do not have the power to force change on a government or on institutions which perceive these changes to be against their own interests. Scientists must first build the ties with the rest of the people necessary for any substantive change. Furthermore, the substitution of a scientific élite to make decisions instead of a political élite holds no greater guarantees for the well-being of mankind. A necessary step then is for scientists to bring science to the people. The press will always tend to exaggerate the significance of scientific findings

until the time when science is not unusual but a part of everyday life. The scientist can no longer afford to have contempt for the people and for their ability to comprehend. For sooner or later he will have to pay a price for his neglect. Scientists must devote some of their energies to education of the public in a way which will reach all people.

An additional component of this educational role is the obligation of the scientific community to respond to the distorted use of scientific methods to support racism and discrimination against women.

4. Finally, and possibly most important, the radical scientist must operate within his place of work on the same principles he uses to speak to the outside world. If it is important to raise the level of communication with people outside the laboratory, it is certainly as important to build greater unity between all workers within the laboratory. This can be done by working to destroy the hierarchical structure of labs, to support the causes of other workers within the institutions, and to attempt to influence the role of one's institution in the outside world. Hopefully it is from beginnings like this on a small scale that the greater changes in society will arise.

In summary, I would like to emphasize a few points. First, the ideas which I have presented leave many perplexing problems unresolved. The question of what kind of research a biologist should be doing who feels strongly on these issues is left open. These are contradictions which do not seem easily resolvable to me.

Second, the problems I have discussed will almost certainly increase for biologists in the near future. The progress in molecular biology and other biological fields has been so spectacular in the last few years, that it would be dangerous not to work on the assumption that this work will be applied extensively in the near future. This will mean not only the possibility for misuse, but also the likelihood of greater ties of biologists with drug companies who are rapidly beginning to see the potential of research in cancer, molecular genetics, immunology, etc.

Finally, I have indicated my objections to the concept of a scientific élite which considers itself best equipped to determine the nature of the control over science. I believe the nature of this meeting reflects this élitist concept. People have been chosen for the most part on credentials – FRS, Nobel Prize, etc. I am chosen because of the publicity we received, despite the fact that everyone realizes that our isolation of a gene had none of the importance attached to it by the press. There have been discussions of greater contact with the public – where is the public – why haven't we started here? Where among the speakers are the younger people who don't have the credentials of scientific accomplishment, but certainly have credentials for the far greater concern they have about the social impact of biology than many of the speakers here? Why is there only one woman speaker on the programme? I hope that in the future, meetings of this sort will bring together more than just the élite of a particular scientific field.

References

1 Editorial (1969). 'On which side are the angels', *Nature 224*, 1241–2.
2 *Ibid*.
3 In a brochure entitled 'Relevance to sponsor' (of the Cambridge Project) – distributed by Harvard University to its faculty.
4 Seikevitz, P. (1970). 'Scientific responsibility', *Nature 227*, 1301–3.

Discussion

Sheridan I should like to thank the speaker for the first clearly understandable contribution from my point of view as an ordinary person. I feel I am the ordinary person who is not consulted, except at election time when I may put a cross. I don't consider this worth very much because the politicians do not speak my language and don't say anything worth while to the working class. I don't know if people saw a programme on television where a woman said that the Chairman of the Housing Committee referred to her as 'scum', and personally I call myself working-class scum because really I have no true democratic rights. I haven't been asked if the pill or an asthma drug or a cancer drug or any drug should be developed.

The questions I want answered are: why can't I have a home, why should I when I was three months pregnant, because I couldn't afford the deposit on a house, be separated from my husband, and why should I, 'working-class scum', when I was in an institution for homeless mothers and children, be separated from my child?

Beckwith When the lady was talking about not being involved in decisions on these things, I thought I could sense that the reaction of at least some here was that we cannot (and certainly I have read many opinions like this) bring people into decision-making on these questions because they are not prepared and do not know enough. This is a problem which we have to face right now and be involved in bringing these problems before the people immediately.

Curwen I feel that ordinary people are much more adult and able to make the right decisions than would appear from the apathy shown by them about big issues, which is partly because they feel very helpless, but chiefly because they do not know the truth. For example, once it was shown that nuclear testing could affect health and also that of future generations, there was world-wide opposition to it leading to the partial test-ban. Unfortunately scientists have not been so active in warning against the dangers of the continuing arms race and consequently the ban has not been made comprehensive.

Perhaps scientists' responsibility is to tell us the truth, not in technical jargon please, but in plain words.

Anon Dr Beckwith, I have here a draft of the scientists' pledge which has been distributed by a small group of militants. Point number two says: 'It is the duty of a socially responsible scientist not to conceal from the public any information about the general nature of his research and about the dangerous uses to which it might be put.' Would you agree that this is one of the duties of a socially responsible scientist?

Beckwith Yes.

Anon If so, what about all those scientists who refuse to accept this principle and work under various official secrets acts which forbid them to publish their research, or to inform the general public about the general nature of their research. Would you agree that they behave in a socially irresponsible manner?

Beckwith Yes, but I think I would also say that many of us who are doing unclassified research which is published and still never reaches the public are irresponsible also for not discussing the implications of our work. I think there are more problems there but I wouldn't single out that group of scientists who were doing secret research.

Nineteen The disestablishment of science

J. Bronowski
Director, Council for Biology in Human Affairs,
Salk Institute, California

On 2 February 1939 my late friend and colleague Leo Szilard sent a letter to Professor J. F. Joliot-Curie in France in which he proposed that atomic physicists should make a voluntary agreement not to publish any new findings on the fission of uranium. It was a crucial time in history. The date was shortly after Munich; Hitler was riding high, and was clearly determined to make war. And just at this threatening moment, results were published which left no doubt that the atom of uranium could be split. Szilard concluded that a chain reaction could be produced and warned Joliot-Curie, 'In certain circumstances this might then lead to the construction of bombs which would be extremely dangerous in general and particularly in the hands of certain governments.'

Far sighted as Leo Szilard was, it is hard to believe that the events he feared could have been stopped even if other scientists had accepted his invitation to silence. The embargo on publication that he tried to improvise was too crude to be realistic, in a field in which discoveries were being made so fast and their implications were so clear and so grave. At best Szilard's scheme could have been a stop-gap with which he might hope to buy a little time.

Yet it is important that a scientist with the standing of Leo Szilard should have put into words the favourite daydream of the bewildered citizen: to call a moratorium on science. It reminds us that scientists also feel helpless in the rush of events, which unseen hands seem constantly to direct towards more and more massive and unpleasant forms of death. We all want to buy time: a time to reflect, without next year's weapons programme already grinning behind our backs.

A proposal to call a moratorium on science, in a strict and

literal sense, would be as unrealistic today as it was thirty years ago. It could be imposed on scientists (who have a living to earn and a career to make) only by taking away their government grants. And no government would agree to do that in the middle of an arms race, when it is hidebound in the concept of negotiating from strength – which means by threat and counter-threat. How nice, how idyllic it would be if governments could agree to suspend weapons research and counter-research, development and counter-development: because if the conditions for such an international agreement existed, peace and the millennium would already be here.

In fact no scientist, and I hope no one who cares for the growth of knowledge, would accept a moratorium even if it were practicable. The tradition of free enquiry and publication has been essential in setting the standard of truth in science: it is already eroded by secrecy in government and industry, and we need to resist any extension of that. On this ground certainly any idea of a literal stand-still in science is wrong-headed, and I do not take it seriously.

But it would be shallow to find in the popular dream of a moratorium nothing more than this literal idea. There is something there that goes deeper, and that is the idea of *a voluntary agreement among scientists themselves.* The layman sees that he will never put an end to misuse by telling scientists what to do: that leads only to the exploitation of science by those who know how to manipulate power. He sighs for a moratorium because he wants science used for good and is convinced that this can come about only by the action of scientists themselves.

If science is to express a conscience, it must come spontaneously out of the community of scientists. But of course this hope poses the crucial questions on which in the end the whole argument hinges. Is science as a discipline capable of inspiring in those who practice it a sense of communal responsibility? Can scientists be moved, as a body, to accept the moral decisions which their key position in this civilization has thrust upon them?

These questions cut deep, and I do not think that any scientist now can sleep in peace by pushing them to the back of his mind. There are two distinct kinds of question here, which engage different parts of his activity and personality. Both are questions of moral conscience; but I will distinguish the first kind by calling them questions of *humanity,* and the second by calling them questions of *integrity.*

The questions of *humanity* concern the stand that a man should take in the perpetual struggle of each nation to outwit the others, chiefly in the ability to make war. Although the scientist is more often drawn into this (as a technician) than his fellow citizens, his moral dilemma is just the same as theirs: he must weigh his patriotism against his sense of a universal humanity. If there is anything special about his being a scientist, it can only be that he is more conscious than others that he belongs to an international community.

The second kind of moral questions, those of *integrity*, derive from the conditions of work which science imposes on those who pursue it. Science is an endless search for truth, and those who devote their lives to it must accept a stringent discipline. For example, they must not be a party to hiding the truth, for any end whatever. There is no distinction between means and ends for them. Science admits no other end than the truth, and therefore it rejects all those devices of expediency by which men who seek power excuse their use of bad means for what they call good ends.

I begin with the questions of humanity. Since so much scientific and technical talent goes into advice and research on war, it is natural that the moral problem that stands largest in our minds (as it did in Szilard's) is still the disavowal of war. I shall be saying later that this is by no means enough; yet it is certainly the first issue before us, and has rightly troubled the public sense of decency for the last twenty-five years.

No man who is able to see himself as others do can approve making war on civilians with atomic bombs, with napalm and orbiting missiles. Yet he also knows that every government that ordered the development of these weapons, anywhere in the world, did so as a duty to its own nation. This is why the man in the street wants scientists to keep these dreadful secrets to themselves when they discover them. He knows that the heads of state have no choice: each of them has been elected to protect the interests of his nation, and to be deaf and stony-hearted to the interests of humanity at large. In a world in which diplomacy still consists of national bargaining, no statesman has a mandate for humanity, and therefore the man in the street turns to scientists in the desperate hope that they will act as keepers of an international conscience.

Some scientists will answer that they have a national duty, too, which is to disclose what they know to the head of state who has been elected to put it to whatever use he judges best. At a time of world war, when the survival of a nation may be at stake, most scientists will act like this; and they will have a better conscience than Klaus Fuchs, who thought himself entitled to do as he chose with scientific secrets. Even in time of peace, there will be some who will put national loyalty first. I think this is a matter of private conscience, and that scientists who feel that their ultimate loyalty is to their nation should follow their conscience by working *directly for their government.*

But what I think is no longer tenable, in the times as they are, is the stance that Robert Oppenheimer tried to maintain, namely to be a technical adviser on weapons on some days and an international conscience on others. The rivalry of nations has now become too bitter, and the choices that it poses for the scientist too appalling, to make it possible to be involved in weapons and war policy and still to claim the right of personal judgment. And that is not merely a matter of professional independence: it comes from a deeper conflict with the morality of nationalism, government and diplomacy.

Nationalism has now distorted the use of science so that it outrages the aspirations of the user. All over the world, from Jordan to Vietnam, and from Nigeria to South America, men carry in their hands the most precise and expensive products of technology: automatic rifles, radar, infra-red glasses, homing rockets, and all the refined machinery of combat and terrorism. Yet the Viet Cong sniper and the Negro infantryman on whom these gifts are heaped has them only to kill. Nothing like this has been given him at home to live with: he has no toilet there and no bath, not a stick of decent furniture and no tools, no medicines and no schooling to speak of. It is bad enough that the world is full of people who live in such misery; and it is an affront to humanity that there should then be pressed upon them as the blessings of technical civilization the beautiful instruments of murder.

It seems plain therefore that unless a scientist believes as a matter of conscience that he owes it to his nation to work directly on war research, he should not accept any indirect part in it. If he chooses war research, he should work in a government department or establishment. But if he abhors the consequences of national war, he should also reject any grant or project that comes to him from a military department, in any country.

This is not an easy counsel in those countries (chiefly the United

The disestablishment of science 237

States and Russia) where most research in the physical sciences, and much outside them, is financed by service departments. For it does not apply only to work on weapons and strategy, but by historical momentum extends over a large area of fundamental and theoretical research.

The support of science from military funds has come about in a haphazard way. In the main, it grew after 1945 out of the crash methods to finance research which had been improvised during the war. At that time, government money in most countries was channelled through the armed services, and it was convenient to go on having the service departments farm out research directly. Since they were not short of money, they often gave grants even to theoretical scientists working in fundamental fields (such as mathematics and psychology) that had no foreseeable bearing on any practical service need.

This pattern of quasi-military support still persists, both in general research grants and in specific project costs, in much of the work done in physics and chemistry, for example. There is a similar pattern in the research financed by the space programme in the United States; again the theoretical and fundamental scientist finds himself at the end of a long pipeline which is not strictly military, and yet whose command of funds derives from a belief by government that the department is a useful outpost of nationalism.

Because of these arrangements, it is convenient for a worker in the physical sciences to draw the money for his research, whatever it is, from one of these quasi-military pipelines. And since his university usually cannot pay for his research, and often (in the United States, for example) cannot survive without grants, he is compelled to go for help where the money is. So it comes about that physical scientists in different countries, who would not willingly work in war research, nevertheless get their funds from government departments whose main business is related to war.

It seems timely to decide now that this is a bad practice, and that grants for non-military research should not come from quasi-military sources. A scientist who accepts money from such a source cannot be blind to the subtle conformity that it imposes on his own conduct and on those who work with him, including his students. The obligations that he silently incurs are dormant, but they are there, and he will be in a quandary whenever a government in trouble decides to make them explicit.

I have already said that most physicists who take money from government departments have nowhere else to go; so that if they were to refuse these grants, they would have to give up their research. This would be a hardship, but it is unlikely that it could last long. For the technical society that we live in cannot afford to let research languish, and will be forced to create new channels for support if the old channels are stopped up by the receivers themselves.

I shall return later to the problem of creating a new organization of support for non-military research. For the problem applies to all sciences, far beyond physics, and before I discuss it I must establish a case for detaching them also from government departments. To do so, I must move on from the inhumanity of war to the moral standing of government influence in general.

The traditional issue of conscience for scientists has been the use of their work to make war more terrible. But the problem that confronts us now is more fundamental and is all-embracing. Scientists can no longer confine their qualms to the uses and abuses to which their discoveries are put – to the development of weapons, or even to the larger implications of an irresponsible technology which distorts our civilization. Instead they are face to face with a choice of conscience between two moralities: the morality of science, and the morality of national and government power.

My view is that *these two moralities are not compatible*. In world affairs, science has always been an enterprise without frontiers, and scientists as a body make up the most successful international community in the world. And I have already shown in my earlier analysis that, in a world of un-United Nations, the public is searching for someone to act for the human race as a whole, and hopes that scientists will do that.

In domestic affairs also, the morality of power was laid down centuries ago by Machiavelli for *The Prince*, and is incompatible with the integrity of science. This is a more subtle and recent issue, which has grown up with the extension of government patronage to cover all branches of science.

A pervasive moral distortion, a readiness to use any means for its own ends, warps the machinery of modern government. The scientist who joins a committee becomes a prisoner of the procedures by which governments everywhere are told only what they want to hear, and tell the public only what they want to have it believe. The machine is enveloped in secrecy, which is called

The disestablishment of science 239

'security' and is used as freely to hoodwink the nation as to protect it. A great apparatus of evasion is constructed, a sort of plastic language without content, which goes by the euphemism of 'the credibility gap'.

The scientist who goes into this jungle of 20th-century government, anywhere in the world, puts himself at a double disadvantage. In the first place, he does not make policy; he does not even help to make it, and most of the time he has no idea what shifts of policy his advice is meant to serve. And in the second and, oddly, the more serious place (for him) he has no control over the way in which what he says in council will be presented to the public. I call this more serious for him, because *public respect for science is built on its intellectual integrity*, and the second-hand statements and the garbled extracts that are attributed to him bring that into disrepute.

Government is an apparatus which exercises power and which is bent on retaining it, and in the 20th century more than ever before it spends its time in trying to perpetuate itself by justifying itself. This cast of mind and of method is flatly at odds with the integrity of science, which consists of two parts. One is the free and total dissemination of knowledge: but since knowledge leads to power, no government is happy with that. The other is that science makes no distinction between means and ends: but since all governments believe that power is good in itself, they will use any means to that end.

An example from Russia shows how damaging the dependence on government favour is for the integrity of science. There the science of biology in general, and genetics in particular, was dominated for thirty years by a second-rate scientist and charlatan, T. D. Lysenko, whose only skill was in exploiting the favours of two successive heads of state. So he was able to falsify biology on a grand scale, to bring up his students in ignorance and deceit and, incidentally, to do lasting damage to Russian agriculture. These are the consequences of the manipulation of science for the sake of political conformity and power. Yet to my mind Lysenko did a greater harm than all these: by being able to silence those who tried to argue with him, he destroyed the trust of other Russian intellectuals in their scientists.

We live in a civilization in which science is no longer a profession like any other. For now the hidden spring of power is knowledge; and more than this, power over our environment grows from discovery. Therefore those whose profession is knowledge and discovery hold a place which is crucial in our societies: crucial in importance and therefore in responsibility. This is true for everyone who follows an intellectual profession; in the sense that I have just described it, ours is an intellectual civilization, and the responsibility of the scientist is a particular case of the moral responsibility which every intellectual must accept. Nevertheless, it is fair to pin the responsibility most squarely on scientists, because their pursuits have for some time had the largest practical influence on our lives, and as a result have made them favoured children in the register of social importance – some would say, spoilt children. So other intellectuals have a right to ask of them, as favoured children, that they accept the moral leadership which their singular status demands. This calls both for a sensitive humanity and a selfless integrity, and it is the second of these that I am now stressing.

The Russian example is a warning that scientists have to renounce the creeping patronage of governments if they want to preserve the integrity of knowledge as a means and an end which thoughtful citizens (including their own students) prize in them. In my view, *there is now a duty laid on scientists to set an incorruptible standard for public morality*. The public has begun to understand that the constant march from one discovery to the next is kept going not by luck and not even by cleverness, but by something in the method of science: an unrelenting independence in the search for truth that pays no attention to received opinion or expediency or political advantage. We have to foster that public understanding, because in time it will work an intellectual revolution even in affairs of state. And meanwhile we as scientists have to act as guardians and as models for the public hope that somewhere there is a moral authority in man which can overcome all obstacles.

These considerations apply as much to those sciences, for example, the biological sciences, whose support comes from branches of government which have no military connections. For the moral issues that face the community of scientists now are no longer to be measured by the simple scale of war or peace. We see the growing involvement year by year of government in science and science in government; and unless we cut that entanglement, we endanger the integrity of all science, and undermine the public trust in it by which I set such store. The silent pressure for conformity exists whenever grants and contracts for research are under the direct control of governments; and then (as the Russian example shows) no science is immune to the infection of politics and the corruption of power.

The time has come to consider how we might bring about a separation, as complete as possible, between science and government in all countries. I call this *the disestablishment of science*, in the same sense in which the churches have been disestablished and have become independent of the state. It may be that disestablishment can be brought about only by the example of some outstanding scientists – as the great Kapitza refused the directions of Stalin in Russia, and Max von Laue refused to work for Hitler in Germany. But the immediate need is more practical. It is to have all scientists consider the form that disestablishment should take, and for which they would be willing to make common cause.

Evidently *the choice of priorities in research should not be left in the hands of governments*. This is a view that government departments will not like, so that scientists who hold it will need to be single-minded if they are to make it heard. They may have to refuse to apply for grants and contracts that are allocated directly by departments. Again, this would be a hardship for many scientists, who now have nowhere else to go for money and who would be forced to suspend their research. But they must be willing to face the hardship for a time if they are serious and united in the will to put science into the hands of scientists.

In the long run, the aim should be to get a single and overall fund or grant for research, to be divided by all the scientists in a country. This would be an effective form of disestablishment, and no doubt governments would accept it rather than watch research become moribund. The alternative would be to let the scientific community build its own fund, and derive an income for itself, by pooling its discoveries and selling the rights to use them.

Once there is a single grant for all research together, its division becomes the business of scientists themselves. They have plenty of practice now in sitting on panels that grade the applicants for the money assigned to each small section of a scientific topic. But in future they will have to undertake to weigh section against section, topic against topic, field against field, and (at the top of this pyramid) each branch of science against the rest.

The disestablishment of science will compel the scientific community to assign its own priorities and divide the overall grant at its disposal accordingly. On the small scale, this is a familiar task and does not seem onerous. The consideration of the work and the promise of other scientists, the criticism of new work and new plans, is an informal part of every scientist's education. At present, it is formalized in the many small review committees which advise on grants in specialized fields. But there is no reason in principle why the procedure should not change its scale, and grow into an overall review of the whole field of scientific research. In most nations, the senior scientific bodies (for example, the Royal Society and the Medical Research Council in England, and the National Academy of Sciences in the United States) make such surveys in piecemeal fashion from time to time. That is only a faint shadow, so far, of a science policy directed in each nation by its scientists. But it is a shadow that points the way to go: it shows that the method is practical, and can be developed from known procedures.

So far, I have proposed two steps in the disestablishment of science: first, refusal to accept grants or projects directly from government agencies, and second, demand for a single national grant which is then to be allocated by the scientific community itself. There remains a third step in the more distant future, and yet it is the crucial step: the allocation of research as a single *international* undertaking.

The public in every nation in the world is looking for an international conscience – that is the point from which I began this essay, and on which it hinges. For it knows that nationalism is an anachronism, and a dying form of civilization. So the public everywhere looks to scientists to find a practical way to express the sense of international duty and decency which is so plainly waiting over the horizon. The reason for this trust is precisely that science is recognized as an international fellowship: international in its principles, and international as a body of men.

In the end, then, the disestablishment of science must mean a change from national to international policies. Evidently, these policies will have to be debated, formed, and put into practice by scientists themselves – our experience of so-called international organizations makes that plain. Alas, we shall not take this third step very soon. True, it has been possible to run international committees on special fields in science already. But on the total scale of science it is an immense responsibility. The body of scientists as a whole will have to develop a system of representation to make its policy, in which the young, the bold, the idealistic and the unorthodox have a better chance to be heard than they do in politics. There is good evidence that this happens in science, and it is the best reason of all for giving scientists their head.

So the scientific community, through its own representatives, will have to judge and balance the importance of the different branches of science at any time, and of the new lines of research in each. It will have to guess the time and the chance of success in each line. And then it will have to combine the judgments of importance with the guesses at success, in order to arrive at a scale of priorities – a scale of claims, as it were – in accordance with which it must divide its overall grant.

It would be nice to believe that the computation of priorities in this way requires nothing more than scientific competence. But alas, even the disestablishment of science cannot make life so simple. There is no judgment of the importance of a field or a line of research that can be confined to its scientific potential. Every judgment in life contains a silent estimate of human and social values too, and the representatives of science will not be able to shirk that. There is no guarantee that scientists will make a better job of fitting science to humanity than has been done so far, but it is time that they faced their moral obligations and tried.

Discussion

Shallice I think we've just had a prize example of liberal clap-trap. There are many people here who think that scientists, because of their position as scientists, can well see that the present capitalist system, and the way in which knowledge is exploited by that system, is going to doom mankind unless something is done about it. If one accepts this form of analysis then the question is not how can we preserve our integrity but how can we best act against this system?

Kafetz We've heard a lot of the possibility that there are some aspects of science which, because they could have dangerous consequences, should not be studied. I would say that if one discovered for example that black people were for various reasons, not to do with environment, less intelligent or less able in some way than white people, then it would be immoral as a scientist to suppress this and it would be immoral to build a society on something you knew as a scientist was not true. Now this may sound liberal clap-trap, but it appears to me that it is far better to spew liberal clap-trap than it is to spew trendy clap-trap, and unless I've misunderstood Dr Beckwith it was trendy clap-trap that he was providing for us.

Nathan The interpretation of politics is reality, and it is reality which has not been introduced into this discussion. I am thinking, in particular, of the ridiculous attitudes of certain people here who refer to themselves in terms of an intellectual élite, or as a conscience. They don't recognize the social reality of the society we live in. When it comes to the power structures and the needs of the power structures in our society, we are all scum; not only those of us who sit here as students and non-scientists, but even the gentlemen in the front wearing their ties and their Nobel prizes and whatever else they may have. And it's this position, this realization of the general oppression of people, of the fact that the social system is not here to satisfy human needs but inhuman needs and that its victims are all of us, whether we are working our forty hours a week, or whether we're fighting for our Nobel prizes and for our mandarin positions in biological departments in new universities up and down the country.

Bronowski No amount of light-hearted abuse, no amount of friendly chat, will divert me from the position that when those of us in this room who chose to become scientists, chose that rather than the profession of bricklayer, they dedicated themselves to something which ennobles science just as there are noble features in the profession of bricklayer. My talk is directed towards what the scientist, as a scientist, should do, not what the scientist as an incidental revolutionary or humanist and the like should do.

Metzger Professor Monod has said that science is bigger than all the scientists. Dr Bronowski said that certain values have to be assumed if science is to go on. We know that some people have related science and religion. None of the statements – direct or indirect – by the eminent scientists has given us any ground for thinking that science is not based on belief; there exists a vast theoretical and physical structure in which scientists are enmeshed. It is an awareness of this condition that has sparked off some of the hostility expressed. But would an increase in social commitments resolve the problems. Certainly these demands are justified, but I doubt that here lies the entire resolution of the crisis. We are faced by issues that are deeper than the political level. The problem also resides in the particular fabric of science and technology – not merely in its social applications. It seems to me – and I am speaking as an artist – that the most challenging and profound, and ultimately the most constructive research activity in science, is that effort to establish new and revolutionary insights into the nature of science and technology as it has developed in different cultures in the past thousands of years.

Rosenhead We've had a debate concerned predominantly with pure science and the academic approach. There are reasons for this, one of which is, of course, that consideration of the social impact of applied science is much more difficult. The conclusions are much harder to arrive at, they are not going to be in any sense, with or without inverted commas, objective, and they are probably going to be controversial. We can't really consider the function of science, the function of the university, or the function of industry, unless we also consider the way in which the social system of which they are a part is functioning.

This meeting has largely tried to maintain a separation between science and society. Contrary to Professor Bronowski, I do not believe that it is the lack of integrity of individuals which counts – so that if we all told the truth the problems would be removed. Again I disagree with Professor Bronowski when he implies that if we could just do away with our Ministry of Defence grants, and our NATO grants, etc., then we would be absolved of responsibility. The point is that we are all implicated; we participate in society, we play our role in it, mostly we do not reject that role. I don't think there is a great deal of advantage in pointing the finger of moral scorn at individuals and saying, 'You should not be doing that' – rather we should be talking about what we can do about the general situation.

And the general situation, as I see it, is that the problem of irresponsibility is the irresponsibility of institutions, not the irresponsibility of individuals. There are institutions in society which are irresponsible because they are intrinsically irresponsible – it is of their nature, it is of their function. If an industrial firm tried to be socially responsible, it would be acting contrary to its own nature. It has been set up for certain ends and it serves those ends and also certain other ones. It serves, however imperfectly, the ends of the shareholders, in terms of profits; it serves, perhaps more meaningfully, the aims of the managers in terms of expanding their area of power. If it tried to act otherwise it would be acting counter to the reasons for which it exists. We cannot expect this particular organism to provide its own cure.

That is my general perspective; I would like to go on from there to ask what we can do next. We have achieved something through discussion in raising the level of all our consciousnesses. What else can we do? One suggestion which has been made is that we should hold a meeting to discuss the problems of industrial irresponsibility. That is one move forward. We will still be talking, but the ideology will necessarily come in explicitly and the consensus will go out, and that is a good thing. It is also an area in which there are no experts so that we cannot have as platform speakers Nobel prizewinners in this area of social responsibility. Maybe this will be an occasion on which more genuine participation could take place.

What else can we do? I would agree with Dr Beckwith that we must look for what we can *do*; not what we can talk about, not what we can analyse, but what we can do, as scientists. There are of course opportunities for action as citizens: we can be politicians or we can be revolutionaries – that is another argument. But what can we do as scientists for the rest of society? And I would say not only *for* but in collaboration *with* the rest of society. I don't want to act with and for the *whole* of society as it now is. Some parts of that society are very well capable of looking after themselves, thank you very much. There are elements in society, however, which suffer most and there are institutions within society which are most directly responsible for the suffering and for the fact that these people lack the possibility of realizing their own potential. I should like to see what scientists can do, with and for those elements of society, and against those institutions.

Twenty Possible ways to rebuild science
M. H. F. Wilkins

Broadly speaking, there were two main points of view expressed at the conference. Scientists trained before the war emphasized the significance of objectivity in science (e.g. Monod), the importance of extending objective approaches throughout our living, and the great difficulty of doing this. Younger scientists (e.g. Young) discussed how social attitudes affect the way science develops and the conclusions it reaches. As was pointed out, these effects are more obvious in some areas of biology than in physics and chemistry. From the discussion it appears that some of the audience felt that scientific objectivity was entirely illusory; but I think it was not the speakers' intention to show that. If there were not a large element of truth in the general idea of scientific objectivity, science could not have developed. It would, I believe, be a serious mistake to over-react to too much emphasis on objectivity in pre-war science, by suggesting that objectivity is wholly an illusion. This matter of objectivity is important because it relates to the possibility of changing science so that it is more adequate to human needs. We have in all these discussions to distinguish between science in general and particular forms that science takes in particular situations. It has been suggested (Bohm) that to fail to make such distinctions leads to fragmented thinking and thus to confusion.

There were other examples of what may appear, as a result of fragmented thinking, to be opposite views. Some scientists, mainly the younger, clearly had great social concern and approached problems in an emotional way, e.g. in discussing the possibility of banning research in certain areas or of ostracizing scientists working on biological weapons. In contrast, some scientists show considerable capacity for careful thought without adequate consideration of human feeling. Radical youth is often criticized for being too emotional and too concerned. To solve our problems we need all the concern we have seen expressed and a lot more; but this will be of little use unless combined with very careful thinking. It will probably be agreed that *really* intelligent thinking *must* involve human concern. As suggested in my introduction, the crux of the problem of 'social responsibility in science' is to find ways of relating thinking and feeling; these being the opposite activities of the mind which, according to Jung, need to be brought together to make man whole. The traditional view of scientists that thinking is superior to feeling is mistaken: both are essential. Critics of science, e.g. Roszak[1] imply that feeling is superior to thinking, that science in general relies too much on the power of reason and logic and neglects moral values. Certainly, one of the main faults of our technocracy is that thinking and feeling are not in a right proportion. But there is not too much thinking, rather there is too little feeling. Moreover, the thinking is itself *inadequate* – it is too restricted, analytical, and lacks comprehensiveness.

One of the main purposes in planning the conference that led to this book was that consideration of the impact of biology would give clearer ideas about the value of science today and in the future. I must admit that these expectations were not fulfilled: few new ideas emerged. The benefits which have been and will be provided in medicine and agriculture were overshadowed by present difficulties created by applying biology and by the way biology has been misused. The mood of the conference, when it was not openly gloomy, was cautious and restrained. It was significant that the word 'optimistic' became a term of reproach. Lack of positive ideas was certainly not due to lack of talent. Talking with younger scientists, I was struck by their almost total lack of positive feeling about the value of science today, or about its potential value if transformed in the future. It would be a mistake to gloss this over by saying that some of the younger people form an insignificant left-wing minority and that the majority of scientists (80 per cent?) still work contentedly. That may be so, however the younger research scientists were more critically minded and more aware of social problems than most of their fellows. The pessimism at the conference indicates that the contentedness of the majority of scientists today is achieved by their working in blinkers, that they do not pay attention to the difficulties outside their daily work. If present trends con-

Possible ways to rebuild science

tinue we may expect that it will be increasingly difficult for society to get the help it needs from scientists.

The main reason for the pessimism is that applying science today produces undesirable effects. Some believe this is primarily the fault of capitalism. Discussion at times seemed to imply that science positively lends itself to misuse under capitalism. Roszak claims that this is true for technocracy in general and that, since technocracy is itself an inevitable product of present-day science, it is the intrinsic nature of all science to produce undesirable effects. It is possible that the audience may also have been given this impression by some of the discussion questioning the 'neutrality' of science, pointing out that science today is misused and has become a repressive tool of governments. Material misapplication in peace and in war, though of the greatest importance and urgency, may in principle be largely avoidable by social and economic planning and by political change. But misapplication cannot be separated from the subtler and more difficult problem of the dehumanizing effects of science. Anti-science opinion today bases its most serious case against science on this dehumanizing. I believe we must face dehumanizing openly and fully; it is at the root of our troubles, and has been too much ignored or denied by scientists. Anti-scientists claim that the practice of objective thinking not only dehumanizes scientists but produces a technocracy which dehumanizes people generally. There is much evidence of this happening today but little to suggest, as anti-scientists appear to claim, that science *necessarily* dehumanizes. It is true that our society has not appreciated the limitations of purely intellectual processes and mistakenly tends to believe that thinking alone can solve our problems. For example, consider the committee of reasonably humane men who had to decide what use to make of the first nuclear bombs. They thought carefully about many possibilities and then recommended the one course of action which we would now probably agree was the worst: to drop bombs, without warning, on concentrations of civilian population. Since committees and experts often do not give the right answers, some technocrats propose, without sufficient understanding of what is involved, to solve these problems by means of computers, thus making machine values dominate human values. Also dehumanization can occur if individual man is regarded solely from the viewpoint of social planning, or of psychology, neurophysiology, molecular biology and so on.

Technocracy, and scientists individually, undervalue the importance of non-intellectual human experience. Yet scientists, like other people, would agree that what is most valuable in life are such things as human relationships, love, happiness, a vague conviction that somehow life is worth living. These aspects of experience contain a large element going beyond intellectual and analytical thought, ideas about them being developed in poetry, art, and related activities. Restriction and under-development of this spiritual and feeling aspect of man is what is really meant by dehumanizing by science.

Those who are against science have been quite rightly disturbed by the consequences of science today. But they have over-reacted sometimes even to the point of irrationality, and have lost sight of the elements common to science and other human activities. Because in science today reason has mainly a limited and special form appropriate to analytical science, they associate reason with narrowing and with elimination of feeling. They come to regard rationality as essentially negative and constraining to freedom of the spirit. This is a common confusion today. Comprehensive reason may in fact be a positive guide to creativity and to extension of experience, for example in music.* The solution of our problems is not to destroy science but to rebuild science in comprehensive form which includes man and human needs as part of the content of scientific enquiry. Such science should help to liberate the spirit rather than to constrain it.

Some ideas about the value of science were presented in the Introduction. This value must lie primarily in the contribution of science to what we may call the growth of man's mind or the extension of his experience. The applications of science contribute to this. In themselves these applications may be very useful, but that is not their primary importance. There is an essential component of reason in the valuable experiences of man in human relationships, in art and in religion. Without a rational side they have no form or value. Art, as well as science, needs discipline: artists think as well as feel. All science consists basically of thinking and checking ideas against observation. Although the scientific method works most easily in science subjects, where data can be well defined, it is also applied in fields of study such as history, literature and so on. There is no clear demarcation between science and the thinking part of human activities

* I am not able here to deal with the relation of reason and feeling. Moreover, the argument is only roughly sketched in.

Possible ways to rebuild science 251

generally. We may regard the whole of our lives as an enquiry in which our ideas are tested against our experience. Even the critics of science would probably agree that man's development must include the rational and thinking side; and from this we cannot exclude the development of science. Therein lies the value of science. We may also note that *true* scientific objectivity requires honesty, freedom from prejudice, consistency and breadth of view – all attributes of traditional value. That is why science has often tended to be associated with humane attitudes and support of freedom. This tendency is today no longer obvious, not only because of the inadequacies of our politics, but because our present science is too narrow and not suited to the problems which, in the wider sense, now face it.

In what ways can we work towards building a more adequate science? First, we can try to get clearer ideas about the nature of science. A scientist tends to work in a fixed framework and it is often very difficult for him even to glimpse the possibility, in his own science particularly, of approaches different from the one he uses successfully. The scientific community resembles a well established Church in having firmly held doctrine. Science today very successfully gives answers to many of the questions it sets itself, e.g. consider the analytical, reductionist approach of molecular biology. The questions science asks itself are mainly of a narrow kind. The narrow approaches suitable for answering such questions are inadequate when applied in broader science which involves the social impact of science. In relation to this let us consider the matter of scientific objectivity. Bohm approaches this by considering the holocyclation[2] of a totality which includes the object, the observer, his general conception of nature and of society, his ideology and other relevant factors of the individual. As an example we have ecology, where solutions are only obtained by considering all the parts, no essential factor, however small, being left out (except that man's conception of nature as an operating factor within the totality is generally ignored).

It is clear that we need much more study of the social and historical (e.g. Young), economic, political and philosophical aspects of science. Similarly, much needs to be done to bring such studies into science education in schools, universities, and into popular presentations of science generally. At some universities very good beginnings have been made, e.g. at Edinburgh, Manchester, Yale. The scientific societies should play their part in broadening the perspectives of science. The report by Hayes that the newly-founded genetics society in Germany would concern itself with the social implications of genetics is indeed significant.

It is important to try to go beyond these current trends in broadening science. Special attention should be given to activities which combine science and art – making new links between thinking and feeling. Art has its own crisis. Some of us in the BSSRS are giving lectures about science to art students. The society is arranging talks on enquiry containing elements of science and of art for science students and is exploring the possibility of courses for children. Clarification about the nature of art/science emerges during these activities. The intention is not to add art as a subject to the education of scientists. That would be mere fragmentation. The purpose of these hybrid studies is to give new understanding of the nature of science and to develop science in new ways so that it becomes more a part of an integral culture.

There is a growing tendency today for artists to regard their work as an enquiry, an open-ended activity which does not aim at producing an object but nevertheless provides a defined statement. Science may also be seen as an open-ended activity producing statements about aspects of nature rather than providing final truth. In rebuilding our culture, science may integrate with art into a broader human enquiry. The objectivity required today involves breadth rather than the present narrowness. Connected with this is the open-endedness of science and of the problems it poses itself. There is a lack of finality and uniqueness in scientific solutions – new problems continue to arise; e.g. Newtonian mechanics and wave mechanics are valid views of aspects of the universe rather than successive and closer approximations to a final truth.

Building new intellectual frameworks should be stimulated by changing the research priorities within science. It will in any case be desirable to put more effort into studying the effects of applying science and less time into 'basic' research, which often tends to be a rather meaningless piling up of knowledge. We may hope that there will develop new hybrid areas of enquiry where man and the rest of the universe are brought into relation. Current environmental studies are one example. We may expect that the intellectual challenge of such science – involving social aspects – will produce a change of scientific fashion so that it turns away from present emphasis on so-called pure science.

One thing that stood out clearly at the conference is that scientists cannot solve the problems by themselves. We need all the help we can get from the rest of society. The problems are jointly those of society and of science. Scientists and non-scientists must learn to work together. I think it would be best for scientists to be humble and admit that dehumanization* is a real problem. Scientists cannot count on help solely from poets, artists, religious thinkers and so on, who have themselves in their own activities become remote from living. Rather we need to go to all sorts of people. To use my words: 'We have got to get science going out to the people. We, as scientists, have common problems with other people. We are all rather helpless. If the bomb goes up, we go up with everyone else.' I think therefore we were right to discuss these matters in public even though we had not beforehand clarified our ideas as much as we might have done. Discussion between scientists tends to remain in a fixed framework. To work in scientists' study groups means remaining in the old framework.

I think Beckwith was right in advocating that scientists should start in their own institutions and try to share their work with their staff as a whole. I have, myself, found it illuminating to give talks to the non-scientists in our laboratory about our research and why we do it. The audience asked for further talks, for discussions and demonstrations. I was dismayed to realize how remote our non-scientific staff were from our scientific work; for example, many had never looked through a microscope. Our scientists are only too pleased to explain their work to non-scientists, yet this explaining has seldom taken place in our laboratory or elsewhere, not so much because it is difficult, but because it has not been part of the general scientific tradition. The exercise of talking about our work to ordinary people, and possibly justifying it to them, makes us think differently and raises questions about the nature of science.

* Going beyond Roszak, Kierkegaard has said: 'knowledge is an attitude, a passion; actually an illicit attitude. For the compulsion to know is just like dipsomania, erotomania, homicidal mania, in producing a character that is out of balance. It is not at all true that the scientist goes after truth. It goes after him. It is something he suffers from.' This extremist, even poetic, statement ignores the humane influence of science but expresses much truth, and apart from the word 'illicit' is not incompatible with science being valuable. To accept it might make science difficult to do, but there is probably no easy way out.

To get clearer ideas about how to change science we need experiments in the form of new research, teaching courses, meetings and other activities. In trying to develop a new, broader science which is related to human needs we shall presumably not get far without political change. We could expect that getting rid of the profit and power motives in capitalism would reduce abuse of science. However, experience in countries where capitalism has been replaced shows that we may expect to repeat, even after political change, many of our mistakes unless we make a fundamental cultural transformation. Such a transformation would necessarily be aimed, not only at solving the crisis in science, but also at solving the crisis in our culture generally. In that way science might again become a force for the improvement of man.*

References

1 Roszak, T. (1970). *The Making of a Counter Culture*. London, Faber.
2 Bohm, D. 1971. 'Fragmentation in science and in society', pp. 22-35 above.

* This was written after the conference. I am grateful to David Bohm, Gustav Metzger and Geoffrey Brown for suggestions.

Contributors to the discussion

Adinolfi, Matteo: Paediatric Research Unit, Guy's Hospital Medical School, London.

Buican, Tudor: Roumanian undergraduate studying biophysics at King's College, London.

Channon, Cyril E.: Lecturer in Science Department, Matlock College of Education.

Chater, K. F.: Microbial geneticist, John Innes Institute, Norwich.

Corbett, J. R.: Fison's Agrochemical Division, Chesterford Park Research Station, nr. Saffron Walden, Essex.

Curwen, Margaret: Honorary Secretary, Women's International League for Peace and Freedom, 12 Woodland Grove, Weybridge, Surrey.

Dauman, Jan: 34 Redcliffe Square, London, S.W.10.

Dewey, D. L.: Cancer Research Campaign, Research Unit in Radiobiology, Mount Vernon Hospital, Northwood, Middx.

Doll, W. R. S.: Regius Professor of Medicine, University of Oxford.

Edge, David: Director, Science Studies Unit, University of Edinburgh.

Garrood, A. C.: Postgraduate student in plant biochemistry, University of Liverpool.

Gibson, I. J.: Senior Lecturer in Genetics, Hatfield Polytechnic.

Giles, K. W.: Department of Botany, Birkbeck College, London.

Gorinsky, Conrad: Department of Biochemistry, St Bartholomew's Hospital Medical College, London.

Gunnell, John: Department of Education, University of Leeds.

Kafetz, Kalman: Student at St Thomas's Hospital Medical School, London.

Kernaghan, Ann: M.R.C. Biophysics Unit, King's College, London.

Lal, Shivaji: Lecturer in Physiology, Chelsea College of Science and Technology, University of London.

Levy, Hyman: Emeritus Professor in Mathematics, University of London.

Masters, Millicent: M.R.C. Molecular Genetics Unit, University of Edinburgh.

256 Contributors to the discussion

Metzger, Gustav: Painter, sculptor and writer; editor of *Page*, Bulletin of the Computer Arts Socitey.

Nathan, Keith: Vanbrugh College, Heslington, Yorkshire.

Parkhouse, R. M. E.: National Institute for Medical Research, Mill Hill, London, N.W.7.

Perutz, M. F.: Laboratory for Molecular Biology, Cambridge.

Pirani, F. A. E.: Professor of Rational Mechanics, King's College, London.

Pontecorvo, G.: Visiting Professor, University of London; Imperial Cancer Research Fund, London, W.C.2.

Raschid, Salman M.: Psychiatrist, King's College, London.

Rosenhead, Jonathan: Lecturer in Operational Research, London School of Economics.

Rothman, Harry: Department of Liberal Studies in Science, University of Manchester.

Salaman, D. F.: School of Biology, University of Sussex.

Schama, Rosalind: 108 Nell Gwynn House, Sloane Avenue, London, S.W.3.

Shallice, Tim: Lecturer in Psychology, University College, London.

Sheridan, Jeanette: 21 Weadle House, Seven Sisters Road, London, N.4.

Silverstone, Allen E.: M.R.C. Molecular Genetics Unit, University of Edinburgh.

Simon, Toby: 3 Wells Rise, London, N.W.8.

Singer, Jack: Rockefeller Foundation – Population Council Fellow in Human Genetics, Kennedy–Galton Centre, Harperbury Hospital, St Albans, Herts.

Solet, Maxwell, Inman Street, Cambridge, Mass., USA.

Sussman, Arthur Jay: Department of Biochemistry, University of Oxford.

Yon, Robert: Chelsea College of Science and Technology, University of London.

For Product Safety Concerns and Information please contact our EU representative GPSR@taylorandfrancis.com Taylor & Francis Verlag GmbH, Kaufingerstraße 24, 80331 München, Germany

Printed and bound by CPI Group (UK) Ltd, Croydon, CR0 4YY
08/06/2025
01896991-0012